Personal property of
Lt John Hey Clark

Please return

JANE'S POCKET BOOK OF MISSILES

JANE'S POCKET BOOK OF MISSILES

New Edition

Edited by Ronald Pretty

MACDONALD AND JANE'S

FIRST PUBLISHED 1975
THIS EDITION FIRST PUBLISHED 1978

ISBN 0354 01070 0 (PVC EDITION)
ISBN 0354 01069 7 (LIBRARY EDITION)

Printed in Great Britain by Netherwood, Dalton and Co. Ltd., Huddersfield

PUBLISHED BY MACDONALD AND JANE'S PUBLISHERS LIMITED
PAULTON HOUSE, 8 SHEPHERDESS WALK, LONDON N1 7LW

CONTENTS

INTRODUCTION

It is gratifying for author and publisher alike when the opportunity comes to produce the second edition of any book. The success of the first edition is implicit in the decision to print the second, but more important is the scope this gives to improve upon the original. In the case of *Jane's Pocket Book of Missiles* such a chance is especially welcome because of the rapid and exciting advances being made in this branch of military science, and which demand extensive and constant research on the part of the amateur or professional observer of the subject if he is not to be overtaken by events.

Since the first edition of this Pocket Book appeared, the Soviet Union has introduced new families of strategic, airborne, tactical and air defence missiles; America has scrapped her ABM system; and the British, French, West German, Italian, Norwegian, and Swedish defence industries all have significant developments (and some cancellations) to be placed on record. To fulfil these functions within the confines of a pocket book, in a fashion that makes for easy use by the reader, is not simple. While the straightforward alphabetical arrangement of missiles adopted for the first edition had the merit of simplicity, with the total number of missile names and designations running into hundreds (many weapons have several alternatives) this approach has a number of limitations. One of these is the scattering of missiles of the same sort in widely separated pages of the book, thereby making comparisons of different examples of one class of weapons – say, Air Defence – a tedious business.

For such reasons, we have re-arranged the entries in this new edition to bring together all missiles of the same basic type, thus forming six distinct sections dealing with, Strategic Weapons, Battlefield Missiles and Rockets, Anti-Tank Missiles, Air Defence, Naval Missile Systems, and Air-Launched Missiles. As a further aid to comparison, each of these sections contains one or more pages of line drawings of the weapons in that category drawn to a common scale.

The text has been revised and rewritten to reflect the latest information obtained for use in *Jane's Weapon Systems*, and to a certain extent this book is a miniaturised version of the missile sections of that widely respected reference work.

R. T. Pretty

STRATEGIC WEAPONS

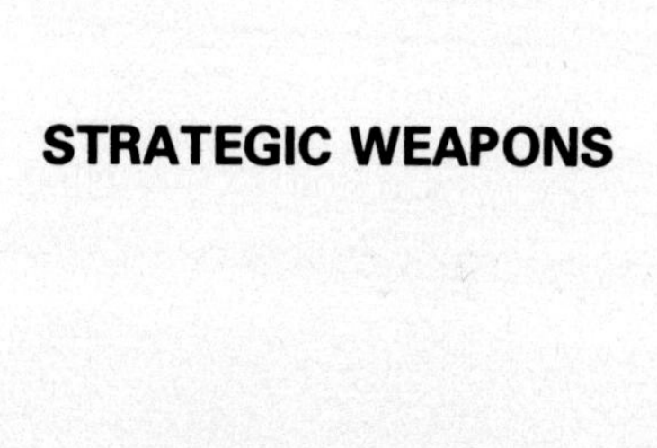

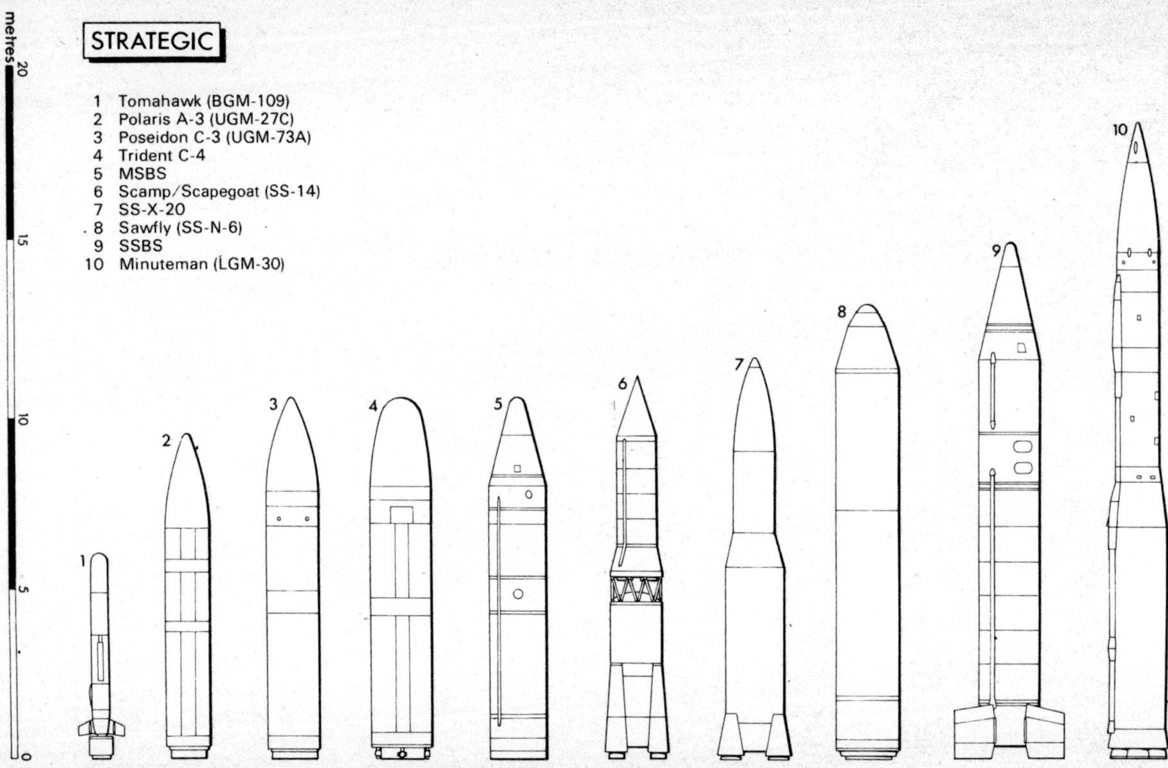
STRATEGIC
1 Tomahawk (BGM-109)
2 Polaris A-3 (UGM-27C)
3 Poseidon C-3 (UGM-73A)
4 Trident C-4
5 MSBS
6 Scamp/Scapegoat (SS-14)
7 SS-X-20
8 Sawfly (SS-N-6)
9 SSBS
10 Minuteman (LGM-30)
metres
20
15
10
5
0
1
2
3
4
5
6
7
8
9
10

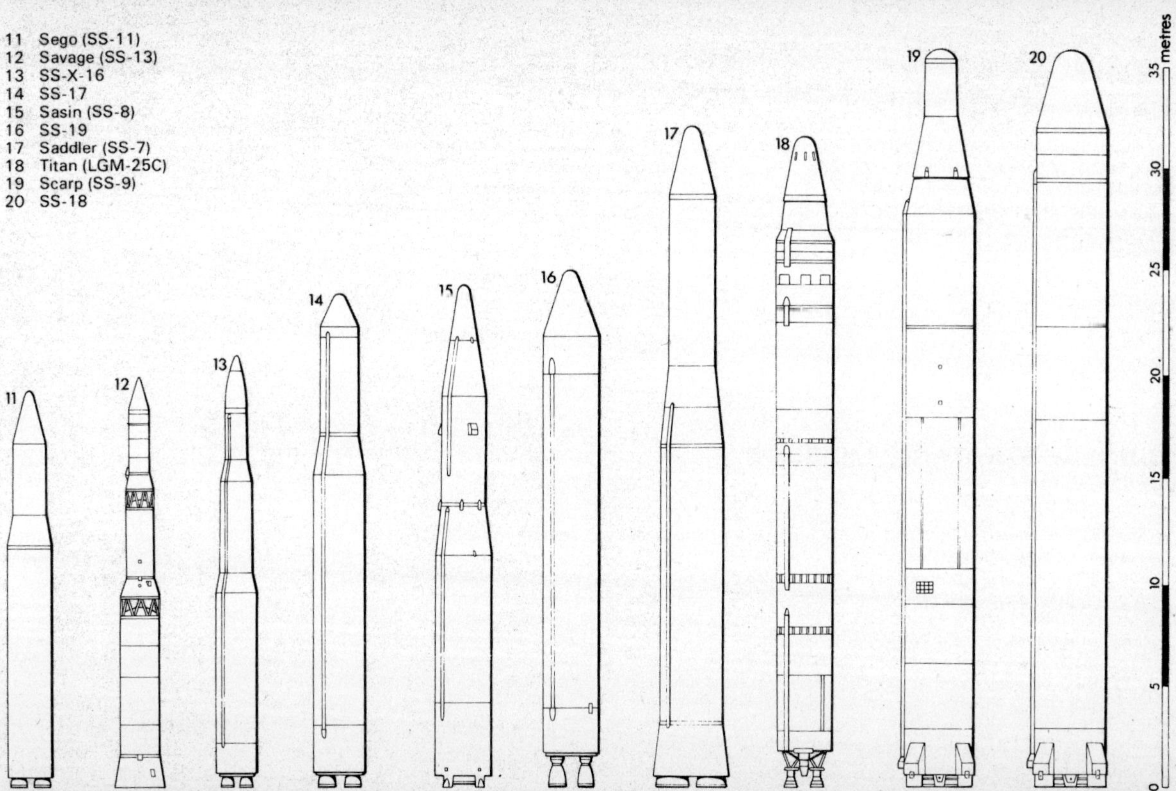
11 Sego (SS-11)
12 Savage (SS-13)
13 SS-X-16
14 SS-17
15 Sasin (SS-8)
16 SS-19
17 Saddler (SS-7)
18 Titan (LGM-25C)
19 Scarp (SS-9)
20 SS-18
11
12
13
14
15
16
17
18
19
20
metres
35
30
25
20
15
10
5
0

STRATEGIC WEAPONS

This section contains information on strategic missiles, most of which fall within the Inter-Continental Ballistic Missile (ICBM) or Submarine-Launched Ballistic Missile (SLBM) categories. Additionally, Intermediate and Medium Range Ballistic Missiles (IRBM and MRBM) have a strategic connotation in certain circumstances, e.g. within the European and Asian continental context, and are therefore included here. The American Tomahawk cruise missile, which differs from all of these by virtue of being non-ballistic, is also included since the strategic role is currently its most significant although a variety of tactical applications are planned for it. However, Tomahawk's American 'running mate', the Air-Launched Cruise Missile (ALCM) and other US and Soviet missiles of a similar type that may have strategic uses, will be found in the pages of this book dealing with air-launched missiles.

CHINESE STRATEGIC MISSILES (CHINA-PEOPLE'S REPUBLIC)

The PRC has made a substantial investment over the past 15 years in research, developmental testing, and production facilities for both liquid and solid propellant missile systems. As the Chinese have a continuing interest in solid propellant rocket motor facilities, and as these are expanding, it is estimated that the PRC will develop solid propellant systems over the longer term.

The Chinese now have a capability for nuclear strikes by missiles and bombers all around the periphery of the PRC, at distances up to 5,900 km. Their capacity to produce fissionable materials is expanding, and a 4MT test, their largest so far, was conducted, during the late 1976/early 1977 period. The Chinese have no present capability to attack the Continental United States directly and are unlikely to obtain one for at least several years. The principal objective appears to consist of providing a modest deterrent to possible Soviet threats.

Soviet reaction to potential Chinese strategic missile threats is implicit in reports that modifications to the Moscow ballistic missile defence system have been made or are planned to adapt it to counter an attack from that direction.

Brief summaries of the salient features of the several categories of Chinese strategic missiles are given in the following entries.

CSS-1 (MRBM)

Liquid-fuelled medium-range ballistic misiles similar to the Russian SS-4 MRBM have been operational in north-eastern and north-western China since 1970, although American reports claim operational use since 1966. Like the SS-4, these missiles have a range of about 1,800 km, and it is understood that some 50 to 70 are currently deployed — the original expectation that the numbers would increase was not realised, and since 1972 the total has remained unaltered. The first operational warheads are believed to be smaller than those carried by the Russian missile; a figure of 20KT has been suggested; but a larger warhead may be fitted to some: in particular it is probable that those which have been most recently deployed carry thermo-nuclear warheads similar to those carried by the CSS-2 IRBM.

CSS-2 (IRBM)

Although the Chinese IRBM and MRBM programmes have been subject to delays similar to those that appear to have affected their ICBM programme, it seems fairly clear that there has been a small-scale (15-20) deployment in China of a single-stage IRBM using a storable liquid propellant. This missile is understood to be similar to the first stage of the rocket which launched the first Chinese satellite in April, 1970 and its range has been estimated as lying between 2,500 and 4,000 km. According to US sources, entry into operational service occurred in 1971.

Basing is understood to be on a basis of deployment at permanent sites.

CSS-3 (ICBM)

The CSS-3 limited-range ICBM system was flight tested in 1976. It is the official DoD assessment that while the PRC has not abandoned the programme, it plans to deploy the system in limited numbers only. Other sources believe the CSS-3 to be deployed in north-west Sinkiang. The estimated range is about 6,500 km, sufficient to give a limited capability to cover targets in Asia, parts of European Russia, Australia, the Middle East, but not the USA.

CSS-X-4 (ICBM)

The CSS-X-4 is China's only full-range ICBM with the potential to hit continental US targets. It is in the same class as the American Titan and Russian SS-9, this implying a range of some 12,000 km, with a 3MT warhead. Under development for a number of years, full flight testing of the CSS-X-4 will call for launches into an open ocean impact area. Analysis of such launches will provide additional information on performance characteristics. A small number could be deployed by the early 1980s.

SLBM

American authorities remain convinced that China intends to develop a two-stage solid propellant submarine-launched ballistic missile, similar in size and capability to the early (A-1) Polaris. At the present time, however, little firm information is available for publication.

MINUTEMAN (LGM-30) (USA)

Of the four versions of Minuteman which have been developed, two are now in service, Minuteman II and III (LGM-30F and G). 1,000 of these missiles (450 Minuteman II and 550 Minuteman III) with 54 Titan ICBMs represent the full strength of America's land-based strategic nuclear force. Both versions are of about the same size but the MK III is heavier. It also is equipped with the MK 12 multiple independently targetable re-entry vehicle (MIRV) warhead, as opposed to Minuteman II's single payload. Both versions are the subject of the Minuteman Force Modernisation Programme that has been in progress for several years and which involves updating electronics and the provision of additional operational facilities. Also included is the development of a new higher yield warhead, the MK 12A. Other development work includes a terminally guided MaRV (Manoeuvring Re-entry Vehicle), and the flight testing of a Minuteman III with a larger number of smaller re-entry vehicles than the three carried in the MK 12 warhead.

Missiles are stored ready for launch in individual hardened launch sites each having a surface area of two to three acres. The individual launch silo is nearly 25 m deep and nearly four metres in diameter with two underground equipment rooms around the silo casing extending more than eight metres below the surface. The flight-launch control centre is buried some 15 m undergound and is a blast-resistant shock-mounted capsule operated by two officers of Strategic Air Command. In 1966 Boeing designed an airborne launch control system which was installed in KC-135 airborne command post aircraft. To counteract the possibility of disruption of these systems by EMP (electro-magnetic pulse) interference resulting from nuclear explosions, guidance hardening modifications have been made and an improved command post arrangement was devised. Known as AABNCP (Advanced Airborne National Command Post), equipment for this system is installed in Boeing 747 aircraft, from which launch commands for Minuteman can be given if necessary.

A total of seven of these aircraft is planned, the first three of which will carry the electronic systems currently installed on EC-135 airborne command posts, to allow early operation of the new aircraft. Later stages of the programme will provide improved communications facilities and data processing equipment.

Another recently-introduced improvement is the Command Data Buffer System. This will enable a single Minuteman III to be retargeted in 36 minutes and the entire force in less than 10 hours. The programme of installing the CDBS should be completed by the end of fiscal year 1977.

In the Minuteman II (single RV), eight targets are stored in the missile's computer, one of which is the primary target. Any of the remaining seven can be selected by a crew-member in a launch control centre. To change the target data stored in the missile computer requires re-programming, which could take 36 hours. All 450 Minuteman IIs are of this type.

The Minuteman III (with two or three RVs) has three target sets stored in its computer, one of which is the primary. Either of the others can be selected by the launch control centre crew. Minuteman IIIs equipped with CDBs have the facility of infinite, rapid, remote targeting. A revised guidance software programme called HEG (Hybrid Explicit Guidance) reduces the targeting information required by the computer. All Minuteman III missiles are being brought to this standard.

In late 1976, funds were allocated for the construction of a further 60 Minuteman III missiles.

Type: Intercontinental ballistic missile
Configuration: Cylindrical body with first stage of larger diameter than second and third stages. No fins or wings. Silo launched
Length:
LGM-30F 18.2 m
LGM-30G 18.2 m
Diameter: 1.83 m max
Weight:
LGM-30F 31 750 kg
LGM-30G 34 475 kg
Propulsion: Three-stage solid
Range:
LGM-30F 11 265 km
LGM-30G 13 000 km
Guidance: Inertial plus on-board pre-programmed digital computer. First stage control by gimballed motors. Four nozzles are also used on third stage of Minuteman II, but the second stage is controlled by fluid injection into a single nozzle. Minuteman III uses latter system for both stages two and three
Warhead: Thermonuclear 1-2 Megaton single warhead or 3X170 kilotons. New payloads in development
Main Contractor: Boeing Aerospace Company

First test flight of USN Trident re-entry vehicles on an Atlas booster rocket from Vandenberg AFB, California

MSBS (FRANCE)

MSBS (Mer-Sol Balistique Stratégique) is a medium-range, two-stage submarine-launched weapon forming a vital part of the French nuclear deterrent force. In general concept it is similar to the American Polaris missile family.

The M-1 version, now being phased out in favour of the M-2, armed the first two submarines of the SNLE (Sous-marines Nucléaire Lanceur d'Engins balistiques) class, each of which is capable of accommodating 16 missiles in two rows of eight launch tubes aft of the 'sail'. Of these submarines, *Le Redoutable* has been in service since 1971 and *Le Terrible* since early 1973. These are the first two of an initial series of SNLE, and *Le Foudroyant, L'Indomptable* and *Le Tonnant* will be in service before the end of 1978. A sixth submarine, *l'Inflexible,* has been ordered and is expected to enter service in the 1980s. Each submarine can launch all 16 missiles in 15 minutes. *Le Redoutable* and *Le Terrible* have both test-fired M-1 missiles under operational conditions.

An improved version of the MSBS has been deployed as a successor to the M-1. Known as the M-2, its principal difference is that of increased range. A new, higher performance, second stage — the Rita II — similar in principle but powered by six metric tons of solid propellant instead of four, replaces the Rita I stage. The new missile equips the third SNLE (*Le Foudroyant*) and the first two submarines will be modified to accept the new missiles during their normal refitting periods. The M-2 will have the same nuclear warhead as the M-1: from 1976 onwards, however, new deployments will be of the M-20 version. This has an up-graded re-entry vehicle system which includes a thermo-nuclear charge, hardening against the effects of an ABM explosion at high altitude, and improved penetration aids.

A completely new generation of MSBS, designated the M-4, is at the pre-definition stage, the range will be significantly increased to about 4,000 km and multiple thermo-nuclear warheads are planned. M-4 missiles will be deployed on existing submarines after certain modifications have been carried out.

The M-4 is expected to be of the same length as currently deployed MSBS, but of larger diameter. Weight is of the order of 35 metric tons and six or seven MIRVs, each of some 150KT yield, will form the payload. A three-stage propulsion system will be employed; the first stage (M-401) will have a metal casing, while the other two motors (M-402 and M-403) will have wound casings, the former of glass fibre and the latter of Kelvar produced by Du Pont in the USA. Dimensions of M-402 are 2 m diameter by 2.5 m, and of the M-403 1.5 m diameter by 1.5 m.

The experimental submarine *Le Gymnote* is being prepared for M-4 tests in the 1978-79 period.

Type: Submarine-launched strategic missile
Configuration: Cylindrical body with conical nose having a rounded tip. No fins or wings
Length: 10.4 m
Diameter: 1.5 m
Weight: 18 000 kg
Propulsion: Two-stage solid. Type 904 first stage, Rita I second stage
Range: 2 500 km Approx.
Guidance: Inertial. Gimballed nozzle control of first stage, thrust vector control of second stage
Warhead: Nuclear. Reported 500 kiloton
Main Contractor: Aérospatiale

(Data are for M-1 model. See text for variations in propulsion, payload, dimensions, etc of M-2, M-20 and M-4 models.)

M-X (Advanced ICBM Technology)

The Advanced ICBM programme, M-X, is aimed at evolving the technology leading to the development of an entirely new ICBM to ensure the USA a realistic option for modernisation of its ICBM forces in the 1980s and after. When the programme was instituted in 1974, three main problem areas were addressed: (1) selection of the preferred basing and deployment arrangements; (2) the particular guidance requirements for mobile missiles (both air-launched and ground-launched); and (3) the technology demanded for more efficient rocket motors.

The resultant weapon will be a large throw-weight, MIRVed system with a gross weight of between 68,000 and 77,000 kg capable of carrying up to eight re-entry vehicles (RVs) in the 160 kg class.

In the US view, by the early or mid-1980s Minuteman silos may have become increasingly vulnerable to the Soviet ICBM forces, which accounts for the interest in new (mobile) basing modes.

Mobile ICBMs are inherently less accurate than fixed base ICBMs so that there is a need for more capable guidance systems to ensure the required delivery accuracies. If the USSR should decide to replace all of its existing ICBMs with the new family of ICBMs, the Soviet ICBM force could achieve a throw-weight advantage of five or six to one, that is about five or six million kilogrammes against the American total of just under one million kilogrammes. This is the spur to development of new rocket motor technology to provide greater throw-weight per pound of propellant.

There are several types of air- and land-mobile options under consideration, and a leading land-based contender is the so-called shelter system. This depends for its survivability on deception, the concept being to have the missiles mounted on transporter-launchers which would move from one relatively hard shelter to another within a complex of multiple shelters. An attacker would have to target all shelters to be assured of disabling the complete M-X force since he would not know those shelters that were occupied and those which were not.

The air-mobile system would be the most expensive to acquire and to operate. It would demand the procurement of a fleet of suitable aircraft, such as the C-5 or Boeing 747, and to ensure survivability the aircraft with its on-board missile, preferably, would be kept on airborne alert.

In view of the many problems that remain to be resolved, the DoD proposed advanced development of an ICBM that could be deployed interchangeably in the existing Minuteman silos, in a land-based shelter or random deployment mode, or in an air-mobile mode. It could be designed to be cold-launched from a canister in a silo or on a transporter-launcher. In the air-mobile system, the missile could be pulled out of the canister by parachute and fired when vertical stability had been achieved. Tests of this technique were successfully carried out using a Minuteman I and a C-5A aircraft in late 1974. Funding in 1977 and 1978 is planned at $69m and $294m, respectively, for research and development.

POLARIS A-3 (UGM-27C) (USA)

The A-3 (UGM-27C) is the latest of the Polaris series of submarine-launched ballistic missile and is the model now in service with the USN and RN. It represents a major re-engineering of the A-2 which it replaced. Range of the Polaris A-3 is 4,630 km, much of this increase over the 2,780 km of the A-2 being attributable to reduced structure and component weight and the use of a new solid-propellant for the two-stage motor. Both stages of the A-3 motor are of glass fibre construction. The inertial navigation system of the A-3 missile is about 60 per cent smaller than that of the A-2. Overall length is increased by 15 cm to 9.55 m, but diameter and approximate weight remain the same at 137 cm and 13,600 kg, respectively.

Initially, the A-3 missile was fitted with a single warhead of about 1 MT — not greatly different from that of the A-2. Subsequently, however, a MRV warhead was developed for the missile; and most of those now in USN service are fitted with a 3 x 200KT MRV warhead.

The Royal Navy's ballistic missile submarines are equipped with the A-3 Polaris missile supplied to the UK without warheads, and nuclear warheads of UK design and manufacture are fitted, together with re-entry bodies, fusing and arming devices.

At present, each missile is capable of a payload comprising three RVs of an estimated 200KT yield each. Unofficial reports say that these are to be replaced by six new RVs of about 40 KT yield which are capable of wider spacing on impact. It is further claimed that it is the intention of using each RV against a separate military target, rather than launching the three larger RVs in a group against soft targets such as a city. The estimated cost of the British Polaris improvement programme has been quoted as £450 million and its duration as four years.

Targeting data for a variety of operational contingencies is carried aboard each submarine on patrol in the form of prepared magnetic tapes. This information is fed into the fire control system, together with navigational data from SINS and sea state information from the ULCER system. The latter enables compensation to be made for currents which could affect the missile after it is ejected from the launcher tube.

Type: Submarine-launched strategic missile
Configuration: Cylindrical body with ogival nose-cone. No fins or wings
Length: 9.45 m
Diameter: 1.37 m
Weight: 15 850 kg
Propulsion: Two-stage solid
Range: 4 630 km
Guidance: Inertial
Warhead: Nuclear. Single MT and 3 x 200 KT MIRV warheads exist. Most missiles in service believed to carry MIRV type. New warhead for RN being developed
Main Contractor: Lockheed Missiles and Space Company

Polaris A-3 submarine-launched ballistic missile (SLBM) (right) *and demonstration launch of Poseidon SLBM* (far right)

POSEIDON C-3 (UGM-73A) (USA)

Poseidon is a two-stage, solid-propellant strategic missile designed for launching from submerged submarines of the US Navy and is the successor to the Polaris A-2 and A-3 weapons. Range of Poseidon is the same as for the Polaris A-3, 4,630 km, but weight is more than doubled at 29,480 kg. Dimensions are correspondingly greater, but the new missiles are capable of deployment in the launch tubes of existing Polaris submarines without major modifications to the ship.

A considerably increased nuclear payload consists of a MIRV warhead which enables a single Poseidon to be used against a number of targets. An improved guidance system is also incorporated which provides double the accuracy, it is claimed, of the earlier weapons.

The fire control system, Mk 88, is produced for the Poseidon system by the General Electric Company. This interfaces with a later version of the Autonetics SINS (Ships Inertial Navigation System), the missile and the launcher. Mk 88 performs target calculations, insertion of data into the guidance system, test and checkout launch order, and sequence control.

The first flight model of Poseidon was launched on 16 August 1968 from Cape Kennedy, Florida. The 14th launch, in December 1969, was the first complete test of the weapon system, including the launcher, control and missile sub-systems. Development flight testing was completed in June 1970 with 14 of 20 launches described as successful by the Navy. Poseidon became operational on 31 March 1971 when the *USS James Madison* deployed on patrol from Charleston, South Carolina. Thirty-one vessels of the Navy's FBM force are fitted with the C-3 Poseidon. The conversion programme began in February 1969.

Type: Submarine-launched strategic missile
Configuration: Cylindrical body with ogival nose-cone. No fins or wings
Length: 10.36 m
Diameter: 1.88 m
Weight: 29 484 kg
Propulsion: Two-stage solid
Range: 4 630 km
Guidance: Inertial. Thrust vector control
Warhead: Nuclear MIRV
Main Contractor: Lockheed Missiles and Space Company

Since those of the numerous Russian strategic missiles for which NATO code-names have been published have been assigned names beginning with the letter 'S', and those for which only the US alpha-numeric designators are known also start off 'SS', the normal arrangement of this book effectively produces what amounts to a Russian strategic sub-section comprising the next few pages. However, because the American sequence of letters and numbers does not correspond exactly with the strict application of alphabetical listing of NATO names, the following list is provided.

SS-4	Sandal
SS-5	Skean
SS-7	Saddler
SS-8	Sasin
SS-9	Scarp
SS-10	Scrag
SS-11	Sego
SS-13	Savage
SS-14	Scamp/Scapegoat
SS-15	Scrooge
SS-X-16	—
SS-17	—
SS-18	—
SS-19	—
SS-X-20	—
SS-N-4	Sark
SS-N-5	Serb
SS-N-6	Sawfly
SS-N-8	—
SS-NX-17	—
SS-NX-18	—

Of the older types, there are about 750 SS-1, SS-4, and SS-5 in the current Soviet inventory, of which the SS-4 is thought to account for about 500, and the SS-5 as many as 100. The latter two types are expected to be superseded by the SS-X-20 in due course. This will represent a major advance in the level of sophistication in the IRBM force and provide the Soviet Union with an enhanced capability to engage targets in Europe, Asia, and the Middle East. The pre-1964 'heavy' ICBMs, SS-7 and SS-8, are deployed at both hard and soft sites, but under the terms of the SALT agreement the dismantling and/or destruction of launchers has been started. This is to permit their replacement by modern SLBMs. The estimated combined strength of the two types in mid-1976 was in the region of 200 launchers.

Less than 300 SS-9 silos were commissioned and some of these have already been converted as launchers for the new generation of ICBMs, and the remainder are in the course of being converted. SS-11 silos are being converted to accept the new SS-17 and SS-19 ICBMs. The 60 SS-13s now deployed are expected to be replaced in the course of 1977 by the SS-X-16.

The operational deployment of three of the four new Soviet ICBM systems (SS-17, SS-18, and SS-19) during 1975 has since been confirmed, and the fourth (SS-X-16) remains apparently in development still but now in an additional, mobile, form as the SS-X-20.

The SS-X-20 IRBM is considered to be a replacement for the SS-4 and SS-5 missiles. It is a mobile system and comprises the first two stages of the SS-X-16 and has a MIRVed payload. A range in excess of 3,000 km can be assumed.

The Soviet SLBM force consists of about 80 ballistic missile submarines, of which the majority (60 or so) are nuclear-powered.

Four missile types are deployed, the SS-N-6 Sawfly and SS-N-8 being the more modern and most extensively deployed. The earlier SS-N-4 Sark and SS-N-5 Serb missiles are deployed on older submarines, but both types are progressively being phased out in lieu of newer weapons. In 1975 the USSR began to move from the 1972 SALT agreed 'baseline' total of 740 SLBM launchers toward the maximum permitted level of 950 and by November 1976 there were 800 or so operational, with more following.

SADDLER (SS-7) (USSR)

With the SS-8 Sasin, the SS-7 Saddler comprises a complementary pair of Soviet ICBMs which are now regarded as coming within the 'older heavy' ICBM category as implicit in the terms of the SALT agreement. US figures issued in 1974 showed that there were then a total of 209 of both types deployed, some in silos and some in semi-hardened sites. The formal designation 'older' means that the SS-7 was operationally deployed prior to 1964. A single warhead of between 5 and 10 megatons is carried. They are now obsolescent and are being scrapped to permit substitution of newer SLBMs as permitted by SALT terms.

Type: Intercontinental ballistic missile
Configuration: Cylindrical body with second stage of smaller diameter than first stage
Length: About 35 m
Diameter: About 3 m
Propulsion: Probably two-stage storable liquid
Range: 10 500 km estimated maximum
Guidance: Probably inertial
Warhead: Probably 5 Megaton thermonuclear

Sandal (SS-4) is a medium-range ballistic missile which is a developed version of the SS-3 (Shyster) — itself an improved version of the German V2 rocket used in the Second World War. It was first shown in public in 1961 and became a standard MRBM in the Soviet armed forces. Depending on the weight of the warhead the missile has a range of between 1,500 and 1,800 km. The complete weapon system comprises about twelve tractor vehicles with special trailers. Some twenty men are required to erect and launch the missile. The guidance system employed was originally radio command — operating on guidance vanes in the efflux nozzles — but a changeover to inertial guidance has taken place.

SS-4 is now reported to be extensively deployed in the Central Asian region of the Soviet Union. Of some 600 MRBM and IRBM believed to be deployed in the USSR about 500 are SS-4s and it is believed that a substantial number of these are targeted on installations in the People's Republic of China.

Progressive replacement by the SS-X-20 (a mobile IRBM, based on the first two stages of the SS-X-16) is expected to begin during 1977 unless technical or political factors intervene to prevent this coming about.

Type: Intermediate range surface-to-surface ballistic missile
Configuration: Cylindrical body, apart from short flared section at rear end and conical nose, of constant diameter. Small cruciform tail fins
Length: 20.8 m approximately
Diameter: 1.6 m approximately
Weight: 27 200 kg approximately
Propulsion: Single-stage liquid sustainer
Range: 1 770 km approximately
Guidance: Inertial, formerly radio command. Control by moving surfaces on fins and vanes in nozzles of rocket
Warhead: Probably optional nuclear or high explosive

SARK (SS-N-4)

The SS-N-4 Sark is generally regarded as the forerunner of what is now an extensive range of Soviet submarine-launched ballistic missiles (SLBM). This strategic SLBM was first revealed to the public in 1959. It was first fitted operationally in modified Zulu Class submarines and subsequent fittings were in Golf Class submarines, followed by the Hotel Class of nuclear submarine. Z- class boats carrried two missiles and the other two classes each were fitted with three launchers. Sark has now been superseded by more modern SLBMS but may nevertheless be retained in service on a very limited basis.

Type: Submarine-launched strategic missile
Configuration: Circular cross-section body, with fore-part composed of two sections of smaller diameter than first stage motor section. Round tipped nose-cone. No fins or wings
Length: 14.5 m estimated
Diameter: 1.75 m estimated
Weight: About 19 000 kg
Propulsion: Probably two-stage solid
Range: About 650 km
Guidance: Presumed inertial
Warhead: Nuclear

SASIN (SS-8) (USSR)

The SS-8 Sasin and the SS-7 Saddler (which see) form a pair of 'older heavy' ICBMs as implied by the terms of the first SALT agreement between the USA and USSR. Between them at one time the two types accounted for 209 of the Soviet ICBM launcher strength, some in silos, some in semi-hardened emplacements. Both are now regarded as obsolescent. Remaining installations are being dismantled to permit substitution of later SLBMs. Storable liquid propellant is used and a payload estimated at between 10 and 15 Megatons can be carried over ranges up to 10,500 km.

Type: Intercontinental ballistic missile
Configuration: Cylindrical body with second stage of smaller diameter than first stage with tapered fairing between stages. No fins
Length: 24.38 m approximately
Diameter: 2.75 m maximum
Propulsion: Two-stage storable liquid
Range: 10 500 km estimated maximum
Guidance: Probably inertial
Warhead: Probably thermonuclear, 5-10 Megaton yield

First seen in public in 1965, Savage is a three-stage solid-propellant ICBM that has been said to be comparable with the US Minuteman.

The three stages are separated by truss structures and each has four nozzles. The two upper stages are believed to be identical with the two stages of the SS-14 Scapegoat missile.

Although in service for some time, it has never been deployed statically on a large scale. So far as is known, moreover, it has never been equipped with other than a single warhead. One of the new generation of Russian strategic missiles is said to resemble the SS-13 closely and this missile (the SS-X-16) has also been reported to be limited at present to a single warhead. According to US reports, 60 of these missiles have been deployed in silos, and its top two stages are firmly believed to have been deployed in a mobile role. It is interesting to note that just as the first two stages of SS-13 are used as the basis of a land-mobile IRBM, the newer missile is also expected to have its first two stages employed in a similar role as the SS-X-20.

Type: Intercontinental ballistic missile. Fixed and mobile
Configuration: Three cylindrical stages of differing diameters, linked by truss construction interstage segments
Length: 20 m overall
Diameters: 1.0 m, 1.4 m and 1.7 m
Propulsion: Three-stage solid
Range: 8 000 to 10 000 km
Guidance: Presumed inertial
Warhead: Nuclear, probably 1 MT

SAWFLY (SS-N-6) (USSR)

Sawfly is at present the most widely deployed Soviet SLBM. Three versions of it have been identified — all said to be operational and thought to be essentially interchangeable. All are liquid-fuelled, about 13 m long, 1.8 m in diameter, and weighing approximately 20 tons at launch.

The Mod 1 is a single warhead weapon with a reputed range of 2,400 km. In October 1972 flight tests began of a modified version which became known as the Mod 2 and this was given an increased range of about 3,000 km by virtue of an improved propulsion system. According to the American authorities the SS-N-6 Mod 2 is now capable of reaching any target in the USA from the 100 fathom contour off the coasts of the USA. The SS-N-6 Mod 3 is the latest of the Sawfly developments and has a range of some 3,000 km, carrying a payload of two or three MRVs. These are not independently targeted.

The SS-N-6 entered service in 1968, the Mod 2 began operational service in 1974, closely followed by the Mod 3 in the following year. Estimated SS-N-6 operational strength is 544 launchers, all on Y-class nuclear submarines. It is likely that Mod 1 missiles are being replaced progressively by Mod 2 and 3 types. A UK MoD statement in early 1976 reported that total production SS-N-6 amounts to about 1,000 missiles. The expectation is that gradual replacement by the SS-N-8 will eventually result in the disappearance of Sawfly from the operational scene.

Type: Submarine-launched strategic missile
Configuration: Cylindrical body with round tipped nose-cone. No fins or wings
Length: 12.8 m
Diameter: 1.75 m
Propulsion: Liquid
Range: Mod 1: 2 400 km
Mod 2 & 3: 2 965 km
Guidance: Presumed inertial
Warhead: Nuclear in Megaton range, or MRV

The Scamp/Scapegoat intermediate-range ballistic missile in its special tracked erector/launch vehicle on parade in Moscow

The Scapegoat, SS-14, two-stage solid-propellant IRBM appears to comprise the top two stages of the SS-13 Savage missile although the warhead section appears to be different from that of the larger missile and there may be other differences. Its length is approximately 10 m overall. When loaded in its container and transported by its tracked erector/launch vehicle it becomes the weapon system known by the NATO code name Scamp. Before firing, the missile, still in its container, is erected by the powerful hydraulic jacking system at the rear of the tracked Scamp vehicle. In the process the cross-braced framework at the rear of the vehicle is lowered to the ground with the missile standing upright on it. The protective case is then opened to free the missile, lowered and closed again leaving the missile standing on its launching platform. Reported to be deployed in eastern regions of the USSR, although the mobile nature of this weapon system renders categoric statements of deployment of little lasting value.

SCARP (SS-9) (USSR)

Probably the best known of all the ballistic missiles in the armoury of the USSR, the SS-9 (NATO code-name Scarp) is a three-stage liquid-propellant ICBM which was first shown in public in Moscow on 7 November 1967. The missile is some 35 m long and 3 m in diameter and apart from the fact that the first stage propulsion system has six nozzles and four vernier nozzles little information is available from open sources concerning the detailed design of the weapon. Four significantly different versions of the missile have been identified and the following details apply mainly to the first two. Installation started in 1965 and by the spring of 1968 about 225 missiles had been deployed. There was then a nine-month period of apparent inactivity after which installation was recommenced, and there were originally 288 SS-9s operational at deployed complexes. Some of these silos have been converted for the more advanced SS-18, and others are in the course of conversion.

It is believed that Scarp is capable of carrying a warhead weighing between 5,000 and 7,000 kg over a distance of some 12,000 km. In both the Mod 1 and the Mod 2 versions the payload is a single large warhead — that of the Mod 1 having a greater yield than that of the Mod 2. The larger warhead is generally believed to have a yield of 20-25 MT.

SS-9 Mod 3 is a version of the basic SS-9 missile which is reported to have been tested both in a depressed trajectory mode and as a Fractional Orbital Bombardment System, but there have been no known tests of this version since August 1971.

The SS-9 Mod 4 is a multiple re-entry vehicle MRV version, and the designation was allocated in 1973 when MRV tests with the Scarp were resumed after a lapse of two years or more. In January 1973, however, a new test of Mod 4 was detected. Once again, three RVs were carried; but these were of a new design and were equipped with parachutes for soft landing and recovery. Further Mod 4 tests were observed during 1973 and it was noted that there had been some improvement in targeting flexibility, suggesting early experiments in the MIRV techniquue. No SS-9 Mod 4 tests were detected in 1974 or subsequently, and there are no known deployments of this version of Scarp.

SCRAG (SS-10)

Scrag (SS-10) is a three-stage liquid propellant ICBM first shown in Moscow in 1965. It was then said to be a sister vehicle of the launch vehicle used for the Vostok and Voshkod spacecraft. The three stages are separated by truss structures with no interstage fairings. Overall length of the missile is some 37 m, including interstages, and the first stage base diameter is nearly 3 m. The first stage has four gimballed nozzles, the second and third stages have single nozzles — that of the second being very large and that of the third quite small. 'Global range' was claimed for this weapon when it was first shown and, having regard to its size, the published estimates of 8,000 km range seem conservative. It is not an operational weapon, but it may have been an intermediate stage in the process towards the SS-18 missile which is the latest in the sequence of very large Russian missiles.

Lowering a Russian strategic ICBM into its silo, showing details of the SS-9 Scarp Rocket motor arrangement

SCROOGE (SS-15)

Scrooge is the NATO name for a strategic missile which, like Scamp, is carried in a container on a tracked transporter/erector/launch vehicle. Unlike the Scamp container which is hinged to open up the Scrooge container is tubular and is assumed to be used to launch the missile. The missile system is referred to as SS-XZ. Scrooge is longer than Scamp and therefore presumably is intended to carry a longer-range missile. The launcher tube is some 20 m long and 2 m in diameter.

There remains the question of the missile inside the container. Clearly it has to be a fairly thin one, which suggests that it probably uses solid propellants and, on the assumption that the Russians are more likely to have adapted an existing missile than to have developed a new one, it is likely that another version of the SS-13 — possibly with a less powerful first stage — is used. This is, however, no more than an informed guess.

Until recently Scrooge was not believed to be an operational system. It now seems, however, that it is: it has been reported as being deployed near the Chinese frontier in Outer Mongolia.

ЛЕНИН

Currently the most widely deployed Russian ICBM, Sego, SS-11 is believed to be a two-stage missile using storable liquid propellants and having a range of some 10,000 km. It is also believed to be of about the same length as the US Minuteman family, but larger in other dimensions, and to carry a comparable payload. None of these statements can be taken as firmly established and the SS-11 has never been identified with any of the missiles exhibited in Moscow parades.

According to US reports, three versions of the SS-11 have been tested and these have been numbered Mod 1, 2, 3 by the US authorities. Mod 1 or Mod 2 is deployed in large numbers. The Mod 2 is thought to have become operational in 1973, although at one time the absence of any evidence of operational testing was taken as indicating cancellation. However, firings in the latter part of 1975 suggested crew training which could imply deployment. Like the Mod 1, a single RV is carried, but in the case of the Mod 2 this is accompanied by penetration aids. It is probable that Mod 1 Sego missiles have been brought up to Mod 2 standard or replaced by the latter model over recent years.

SS-11 Mod 3 is a comparatively recently-deployed version of the SS-11 two-stage inter-continental ballistic missile which differs from the Mod 1/2 in the nature of its re-entry vehicle and warhead arrangements. It is equipped with a multiple re-entry vehicle (MRV) arrangement of three warheads. The Mod 1 version of the missile, with its single warhead, is said by the US authorities to be considerably less accurate than the US Minuteman missiles but as mentioned above, it is likely that Mod 1 missiles have been replaced by Mod 2s or brought up to equivalent standard.

A substantial number of Sego silos have been modernised and it is thought that a mixture of Mod 2 and Mod 3 SS-11s is now deployed. What the long-term plans for these installations are is a matter for conjecture, but the SS-11 force is expected to decline over the next five to ten years. The pace of the conversion programme of Sego silos to accept SS-17 and/or SS-19 ICBMs is not running as swiftly as was originally expected, and one possibility is that yet another model of the SS-11 may be deployed in future.

Sego (SS-11) ICBM transporters and containers for reloading missile silos on parade in Red Square

SERB (SS-N-5)

Serb is the immediate successor to Sark in the Soviet SLBM family. Though smaller it is thought to offer significant improvements in performance over its predecessor. A single warhead in the Megaton range is probable. When first seen in November 1967 the cluster of 18 smaller, electrically-fired cold-gas nozzles at the base of the missile aroused considerable interest and proved to be a different technique of ejecting the missile from its launch tube. Upon reaching a safe distance from the launching submarine this rocket cluster is jettisoned before booster ignition occurs. Serb is operational with later Golf Class conventional submarines and the later Hotel Class nuclear boats. Each vessel carries either two or three Serbs.

Type: Submarine-launched strategic missile
Configuration: Circular cross-section body, with fore sections of smaller diameter than first stage motor section. Rounded nose-cone. No fins or wings
Length: 10.7 m Estimated
Diameter: 1.5 m Estimated
Propulsion: Probably two-stage solid but liquid has been suggested
Range: 1 200 km Estimated
Guidance: Presumed inertial
Warhead: Nuclear in Megaton range

SKEAN (SS-5) (USSR)

Successor to the Shyster and Sandal liquid-propellant MRBM weapons, the SS-5 is an intermediate range missile also known by the NATO code-name Skean. Although similar in general configuration to its predecessors it can be identified by the absence of tail fins and its blunted nose cone. Control is by means of vanes in the motor efflux. It is carried on a different trailer, towed by the latest type of heavy tractor vehicle and has been shown inside silo underground launch facilities in official Soviet films. The missile was first displayed in 1964. About 100 of these missiles are believed to be deployed in IRBM fields. Expected to be replaced by the SS-X-20 after 1977.

Type: Intermediate range ballistic missile
Configuration: Cylindrical body, slender conical nose, slightly flared skirt at base
Length: About 25 m
Diameter: About 2.4 m
Propulsion: Single-stage liquid
Range: 3 500 km Estimated
Warhead: Nuclear about 1 MT

SS-17

This two-stage storable liquid-propellant ICBM is one of two new Russian missiles regarded as successors to the widely-deployed Sego SS-11 missile. Both it and the other SS-11 replacement, the SS-19, have a MIRV payload, improved accuracy, increased throw-weight and greater survivability than their predecessor. The SS-17 is slightly longer than the SS-11 and of increased volume. The cold-launch technique is used, whereby the missile is ejected from the silo by a gas generator before the main booster is ignited, whereas the SS-19 relies upon the conventional hot-launch technique.

The most important aspect of these new missile developments, however, is the introduction of MIRV-type warheads. Both the SS-17 and the SS-19 are reputed to have on-board computers and have been tested with such warheads; furthermore, since the SS-17 is credited with only a four-warhead MIRV capability whereas the SS-19 is said to be able to carry six MIRV warheads, it would seem that two different MIRV systems are involved. The SS-17 has been tested with a single, large RV which might be allied to the potential accuracy enhancement conferred by the post-boost vehicle to achieve an improved hard target capability.

Tests were observed for the first time during the second half of 1972 and there was intensive testing during 1973 and early 1974. Single RV tests began in 1976. Deployment in converted and modernised SS-11 silos began in 1975, but the rate of conversion has not been as high as was expected initially.

SS-18

Largest of the new series of Russian intercontinental ballistic missiles, the SS-18 is the functional successor to the SS-9. It is a very large two-stage liquid-propellant missile and carries a bus-type MIRV system with an on-board digital computer. This new post-boost vehicle is thought to be similar to those employed in Minuteman III and Poseidon and is capable of dispensing six to eight independently-targeted warheads. According to drawings issued by the American DoD, the SS-18 is of virtually the same size as the SS-9, and the cold-launch technique is employed. Three versions have been identified: The Mod 1 has a single warhead of about 25MT yield; the Mod 2 carries up to eight 1-2MT warheads; and the Mod 3 is another single RV model but rather lighter and more accurate than the Mod 1.

The Mod 1 was the only version on which testing and crew training had been completed by the time that the first SS-9 silos were converted and ready for the SS-18, in 1975, and it is thought that these silos all house Mod 1s. The first crew training launch of the Mod 3 took place in February 1976, and as this version has a maximum operational range greater than the other two models it is expected that more Mod 3 missiles than Mod 1s will be present in the eventual mixture of SS-18 missiles deployed. Alternative payload options for the SS-18 Mod 2 are continuing tests and this version is expected to predominate eventually, possibly with a mixture of payloads.

SS-19

The SS-19 is a two-stage liquid-propellant weapon with an onboard computer and MIRV warhead system. With the cold-launch SS-17, the hot-launch SS-19 has been designed as a successor to the SS-11 missile and has a throw-weight three or four times that of the older missile. It is also of greater volume. Its guidance system is said to combine (inertial?) navigation with a refinement of the traditional Soviet 'fly-the-wire' guidance technique. An on-board computer determines deviation from the pre-programmed course and directs correction to that course or plots a new course depending upon the attendant circumstances. Slightly larger than the SS-17, it is also provided with a MIRV capability, reportedly being able to dispense six RVs. In 1977 it was learned that the Soviet Union was testing a single RV model of the SS-19.

Deployed operationally in 1975, over 100 had been deployed by the end of 1976. The first SS-19 test detected took place in April 1973. With the SS-17, this missile is progressively replacing SS-11 ICBMs, the silos being improved at the same time as they are converted to receive the new type of missile. There are estimated to be in excess of 1,000 SS-11 launchers but we have no idea as to the eventual ratio between SS-17 and SS-19 missiles when the conversion from SS-11 is completed.

SS-N-8

The SS-N-8 SLBM is the latest operational missile of this type in the Soviet Fleet and is deployed in 12-tube Delta I and 16-tube Delta II class submarines. It also has a greater range (7,800 km) than any other SLBM in the world, and can cover targets anywhere in the USA from launch positions in the Barents Sea. Its stellar-inertial guidance is reported to yield accuracies of 400 m CEP. The current model in service is reported to carry a single warhead of 1-2MT yield, but a triple MRV payload is thought to have been developed also. A MIRV payload is another future possibility.

The Soviet force of Delta-class submarines amounts to over 12 boats; six more Delta II boats are building and when added facilities at the Severomorsk yard are finished, an annual production rate of better than 10 will be possible. The SS-N-8 first entered service in 1973.

SS-NX-17 and SS-NX-18

The SS-NX-17 and SS-NX-18 are two new-generation Soviet SLBM in development but so far not deployed operationally. The SS-NX-17 is the first Russian SLBM to use a solid propellant and equipped with a post-boost vehicle (PBV) for delivery of multiple re-entry vehicles (RVs). It began testing in 1975 and is understood to be almost ready for the at-sea stage of further tests. These are likely to be carried out using a converted Y-class submarine as the launch platform. The SS-NX-17 may prove to have a MIRV capability.

The SS-NX-18 began flight test later in the same year as the SS-NX-17 It is similar in some respects to the SS-N-8, but is believed to be larger and equipped with a more sophisticated guidance system. Three MIRV warheads can probably be delivered, and a range capability of some 9,500 km according to US estimates. Prior to November 1976 the SS-NX-18 was tested from land-based launch sites in the USSR, but in early November the first submarine launch took place in the White Sea.

SS-X-16

The SS-X-16 is a new, 'light' ICBM which is seen as a likely successor to the SS-13 Savage. The solid-propellant, three-stage SS-X-16 is reported to have a much greater throw-weight potential than the SS-13 and official US statements say it has an advanced navigation guidance system and a post-boost vehicle of the sort usually associated with dispensing MIRVs.

From the sparse details revealed the SS-X-16 is of similar appearance and dimensions the SS-13. There is insufficient evidence at the time of writing upon which to make any assumptions as to the form of deployment, or indeed whether operational deployment will occur or not. Similar uncertainty surrounds the likelihood or otherwise of deploying MIRVs with the SS-X-16. Current opinion is that while the post-boost vehicle facility implies this capability, testing to date is too limited to suggest imminent operational deployment of a MIRVed version. At the time this edition was consigned to the printers, no MIRV flight test of the SS-X-16 have been observed.

As surmised in the 1976 edition of *Jane's Weapons Systems*, the first two stages of SS-X-16 have been used as the basis of a new land-mobile IRBM, the SS-X-20. A few tests were recorded in 1972 and several more in 1973. In December 1974 two SS-X-16s were launched into the Pacific area to achieve a range of about 8,000 km.

SS-X-20 (USSR)

The SS-X-20 employs the two upper stages of the three-stage SS-X-16 mobile ICBM, and to that extent echoes the relationship between two older Russian missiles, the SS-13 and SS-14. Similar deployment arrangements might be expected, but at present Soviet intentions regarding the SS-X-16 and SS-X-20 are far from clear. A particular US concern is that if a relatively large force of mobile SS-X-20 IRBMs were to be deployed, they could be quickly converted into a powerful mobile force of SS-X-16 ICBMs by the expedient of adding the booster stage to the SS-X-20's two stages. All stages use solid propellant.

Three versions are understood to have been flight tested: one with a single warhead of about 1.5MT and a range of some 5-6,000 km; another with a single, lightweight warhead and a range of between 7,000 and 8,000 km; and a third with a triple MRV payload. In early 1977, the American DoD considered SS-X-20 deployment likely to begin in the near future.

SSBS (Sol-Sol-Ballistique-Stratégique) is a medium-range two-stage solid-propellant missile, with nuclear warhead, which is stored in and launched from an underground silo. It is maintained in a state of readiness, and preparation for firing and actual firing are automatic without human intervention in the launch area. Shortly after the firing order is given the silo door is ejected, permitting launching of the missile. Launch areas are dispersed and hardened to reduce the effects of an enemy attack. Each includes the silo in which the missile is maintained in operational readiness and an annexe housing the automatic launching equipment and servo mechanisms.

The present production version of the missile is known as the S-2 and is the culmination of study research and development programmes that can be traced back to 1959, first experimental launches having taken place in 1965/66. Two groups of nine launch areas have now been established on the Plateau d'Albion. Each group is commanded by a heavily protected, subterranean Central Fire Control Room. These are linked to Strategic Air Force Headquarters by attack-proof communications networks. The first of these groups became operational in the summer of 1971 and the second in 1972. A plan to construct a third group of nine silos was suspended for economy reasons in December 1974.

In 1973 a new French IRBM programme was initiated to develop the second-generation SSBS S-3 Weapon System. This programme also entails the renovation of the first two groups of S-2 silos. Retaining the same first stage as the S2, the S3 will have a second stage of higher performance, namely, the P-6 originally developed for the MSBS. An advanced re-entry vehicle system will also be included. This will have a hardened thermonuclear charge and a re-entry vehicle which is hardened also against the effects of a high altitude nuclear explosion from an ABM. Other features will be a new system of penetration aids designed to counter enemy defences, and a payload fairing to protect the RV during powered flight. Improvements to the silos will include both equipment modernisation and system modifications to increase reliability and reduce maintenance costs. Deployment of S3 missiles is planned to start in 1980.

Type: Medium range surface-to-surface ballistic missile
Configuration: Constant diameter cylindrical body with warhead and re-entry vehicle of smaller diameter carried at fore end. Silo launched
Length: 14.8 m
Diameter: 1.5 m
Weight: Approximately 31 750 kg
Propulsion: Two-stage solid. SEP Type 902 first stage of about 55 000 kg thrust and with four gimballed nozzles. SEP Type 903 second stage of about 45 000 kg thrust and four gimballed nozzles in S2; SEP Rita II (P6) 32000 kg thrust and four freon injectors for thrust vector control in single nozzle in S3
Range: 3 000 km max
Guidance: Inertial. Gimballed thrust motors used for control
Warhead: Nuclear. 150 kilotons (S2): Thermonuclear 1.2 m (S3)
Main Contractor: Division Systèmes Balistiques et Spatiaux of Aérospatiale (SNIAS)

TITAN (LGM-25C) (USA)

The LGM-25C Titan II is an improved version of the earlier HGM-25A Titan 1 ICBM. It carries the largest of all Americann ICBM payloads and has a launch reaction time of one minute from its fully hardened underground silo. The Strategic Air Command of the USAF has three wings of 18 missiles each deployed at sites in Arizona, Kansas, and Arkansas. Several years ago the then US Secretary of Defence suggested that a gradual run-down of the Titan force could be expected, but subsequent events have not confirmed this. On balance it appears that for the present it is being maintained at its level of 54 ICBMs. In November 1976 the USAF took up the option on production of Universal Space Guidance Systems for use in a Titan II improvement programme called Rivet Hawk. Delco Electronics is to provide 25 of these systems.

Type: Intercontinental ballistic missile
Configuration: Cylindrical body of constant diameter. No fins. Silo launched
Length: 31.4 m
Diameter: 3.05 m
Weight: 149 690 kg
Propulsion: Two-stage liquid. LR87 storable propellant first stage of 195 000 kg thrust. LR91 storable propellant second stage of 45 360 kg thrust
Range: 10 140 km
Guidance: Inertial. Control by gimballed nozzles
Warhead: Thermonuclear plus comprehensive penetration aids. 5 ' Megatons yield
Main Contractor: Martin Marietta Corporation

Experimental flight of US Navy Tomahawk cruise missile

This system, also known as the Sea-Launched Cruise Missile (SLCM) because of the intention of providing for surface-ship launching and possibly air launching, is a project to provide US submarines with an underwater-launched cruise missile for both strategic and tactical purposes. The possibility of its use in a land-based form for coastal defence or as a mobile battlefield support weapon is also under consideration. The programme is paralleled by the ALCM (Air-Launched Cruise Missile) project of the USAF and both services have been instructed to ensure that there is maximum commonality between the two weapons.

Both tactical and strategic versions of Tomahawk are to be suitable for launching from standard size torpedo tubes. The tactical model will have a range of 555 km or more and is likely to be armed with a conventional warhead for anti-ship, and possibly shore bombardment, missions. The present warhead design incorporates the Bullpup B warhead, which is already in the US inventory, and is of sufficient size (454 kg) to provide a high single-shot kill probability. The strategic version's maximum range is to be about 3,700 km and a nuclear warhead will be available. In both instances cruising speed will be subsonic, in the interests of range, and at low level for optimum penetration capability.

The tactical version of Tomahawk will use many sub-systems of

the Harpoon although the former will have a heavier warhead and a much greater stand-off range. Harpoon sub-systems directly usable in Tomahawk with minimal changes are the seeker head, radar altimeter, midcourse guidance package and turbine engine. After launch, Tomahawk will descend to a height of a few metres for its low-level cruise to the target area under inertial guidance. A pre-programmed climb will bring it to an altitude at which the seeker can acquire the target and verify its position, after which a lateral manoeuvre and programmed descent is initiated. Subsequently re-acquisition of the target occurs and the missile climbs again to make a diving attack.

The strategic version will follow a low-level, terrain-following type of flight path under an inertial navigation system updated by a Terrain Contour Matching (TERCOM) system. The latter matches the measured terrain being flown over with a digital map of the area derived from existing intelligence gathering sources. The target data and all the routes are stored on magnetic tapes aboard the submarine carrying Tomahawk, and there are provisions for reading this data into missiles before launch. The combined inertial and TERCOM navigation system is sometimes known as TAINS for TERCOM-aided Inertial Navigation System.

A series of successful tests involving about 14 flights was conducted in the USA during 1976 and in February 1977 the two prime contractors, General Dynamics and McDonnell Douglas Astronautics were awarded initial three-year full scale engineering development contracts amounting to $25.1 million. The total USN budget for Tomahawk in 1978 is over $230 million. The US Navy is also looking for a company to design a system which would integrate several types of sensors to detect, classify and select targets for Tomahawk and Harpoon cruise missiles. These would include airborne, shipboard, and satellite sensors and the project is called OTH-DC&T, over-the-horizon detection, classification and targeting.

Type: Submarine-, surface- and air-launched cruise missile. Strategic and tactical variants planned
Configuration: Incorporates provisions for torpedo-tube launching, such as folding wings and control surfaces
Length: 5.56 m, 6.25 m with booster
Diameter: 53.3 cm. Max
Span (Max): Not specified
Weight: Not specified
Propulsion: Air-breathing turbofan, Williams Research; solid booster, Atlantic Research
Range: Approx 555 km - tactical. 2 780 km - strategic
Guidance: TAINS: TERCOM-aided inertial system
Warhead: Alternative high-explosive and nuclear versions
Main Contractor: General Dynamics Corporation

TRIDENT C-4

The Trident I (C-4) missile is a three-stage ballistic missile powered by advanced-technology solid-fuel rockets and guided by a self-contained stellar-inertial guidance system, with a range approximately twice that of Poseidon, or about 7,000 km. The development flight test programme will include from 25 to 30 tests, of which approximately 20 will be development missiles flown from the Eastern Test Range into calibrated impact areas.

The Mk 5 guidance system for the C-4 missile, is smaller and lighter than that for Poseidon, thereby reducing inert weight and providing more space for missile propulsion. It is basically a functional equivalent to the all inertial Mk 3 Poseidon system, the most significant difference is the addition of stellar fixing facilities which enables the Trident missile to meet Poseidon accuracy objectives at longer range.

The Fire Control Subsystem prepares the missile guidance system with targeting and launch position data and co-ordinates the preparations of the launcher and missile test and readiness equipment for the missile launch. Two basic fire control configurations are being developed, the Mk 88 Mod 2 for retro-fitting to Poseidon, and the Mk 98 Mod 0 for the new Trident submarine. The Trident navigation subsystem is basically similar in design to that of the 640-class Poseidon SSBNs with the following improvements: (1) the addition of an electrostatically supported gyro monitor (ESGM) to lengthen the time between navigation fixes and thereby reduce submarine exposure to potential detection; (2) provision for additional closed-loop air cooling to enhance maintenance and reliability; (3) selected modifications to reduce noise, harmonics and susceptibility to stray electromagnetic interference; and (4) re-engineering of the navigation satellite receiver and the navigation sonar system to ensure economic manufacture.

The operational systems development programme contract for the Trident I (C-4) missile was awarded to Lockheed in August 1974. This contract includes missile research and development as well as the initial increment of production missiles. A new Mk 4 re-entry vehicle is being developed for Trident I, and by March 1976 six flight tests of this RV on Atlas and Minuteman boosters had been made. In addition to the Mk 4 ballistic RV for the Trident I, the USN is developing the Mk 500 MARV which will be capable of performing preselected evasive manoeuvres during atmosphere re-entry. It is planned to carry through this programme to the flight test stage to ensure that the Mk 500 will work on the Trident I, but at present there are no plans to deploy it unless some Soviet action should appear to warrant such a course.

The first launch of a Trident C-5 missile was made from Cape Canaveral on January 18, 1977, from Pad 25. It carried an instrumentation payload a distance of about 6,500 km.

Preliminary studies have been made of an improved missile designated Trident II (D-5) but no firm development plans exist.

Type: Submarine-launched strategic ballistic missile
Configuration: Not yet defined, but compatible with Poseidon C-3 launch tube. Dimensions of Poseidon are given below
Length: 10.36 m
Diameter: 1.88 m
Weight: 29 484 kg plus. Estimated
Propulsion: Solid
Range: 7 412 km plus
Guidance: Stellar-inertial
Warhead: Nuclear MIRV, possibly MARV
Main Contractor: Lockheed Missiles and Space Company

BATTLEFIELD SUPPORT WEAPONS

BATTLEFIELD SUPPORT WEAPONS

This section is concerned primarily with large land-mobile tactical guided missiles designed for use as super-heavy artillery over ranges from a few tens to several hundreds of kilometres and many of them are equipped with nuclear warheads.

Such missiles, however, are not the only land-based tactical nuclear weapons available to the commanders of the armed forces of major military powers: conversely, the guided missiles themselves are not necessarily equipped with nuclear warheads. A few brief notes on the main categories of land-based tactical nuclear weapons may help the reader to identify the weapons of interest to him.

Tactical Ballistic Missiles — Guided

These are all substantial ramp-launched rocket-propelled weapons almost all of which have been designed from the outset to be armed with nuclear warheads but most of which have an alternative HE warhead. They are sometimes referred to as short-range ballistic missiles (SRBM) to distinguish them from the longer-range strategic missiles.

Most of these weapons are inertially guided: but the early versions of the Russian Scud missiles are believed to have used radio command guidance and it is believed that one of the missiles developed in Egypt was wire-guided.

Almost all the missiles of this type that are known to have been developed are either American or Russian: the only other weapon that is known to have been completely developed is the French Pluton. A British development programme, for a missile known as Blue Water, was cancelled — but on political/economic grounds: technically it was considered by some to be superior to the American Sergeant missile, itself now obsolescent. The German-aided missile development programme in Egypt never resulted in significant deployment of operational missiles. The situation regarding the Israeli Jericho missile has received little clarification and it now seems that a battlefield support missile — not necessarily called Jericho — does or did exist in Israel. There is, however, still no official confirmation of this. We have received no information on Chinese People's Republic developments in this field but clearly they could develop weapons of this kind if they wished to do so. In Taiwan a medium-range missile known as Coral has been under development at the Chung Shan Institute of Science and Technology for several years.

Tactical Ballistic Missiles — Unguided

Missiles of this type, though still quite widely deployed, must be regarded as obsolescent at least. Weapons still in use are the American Honest John and most of the Russian Frog missiles.

Heavy Tube Artillery

Both the USA and the USSR have operational large-calibre artillery weapons which can fire nuclear ammunition with kiloton-range yields. The American weapons — such as the M-110 203 mm SP howitzer and the M-109 155 mm SP howitzer and variants — are in service in several other NATO armies and a few non-NATO armies. Few, if any, of the weapons in non-American forces have nuclear ammunition available to them, however. Ammunition for the Russian M-55 203 mm gun-howitzer is believed to be controlled in much the same way as is the nuclear ammunition for the American weapons. Some of the M-55 weapons, however, were supplied to the Chinese People's Republic some years ago.

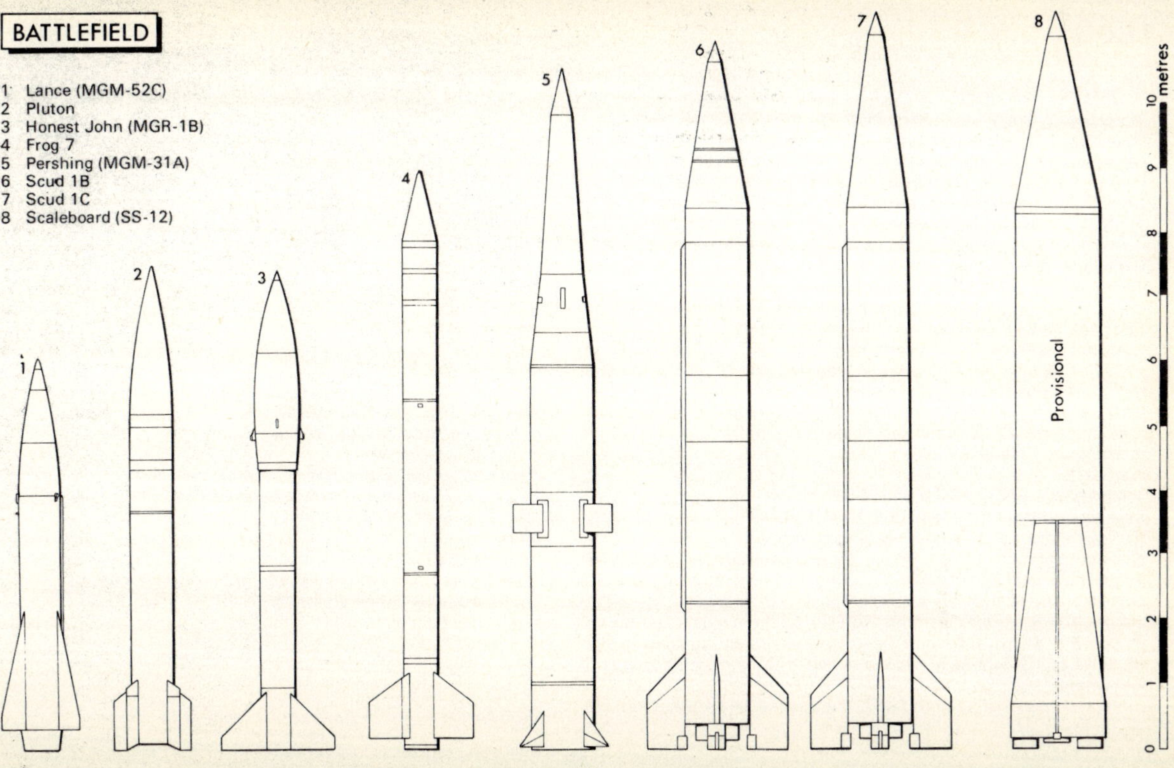

BATTLEFIELD
1 Lance (MGM-52C)
2 Pluton
3 Honest John (MGR-1B)
4 Frog 7
5 Pershing (MGM-31A)
6 Scud 1B
7 Scud 1C
8 Scaleboard (SS-12)
1
2
3
4
5
6
7
8
Provisional
0
1
2
3
4
5
6
7
8
9
10
metres

FROG (USSR)

FROG is an acronym (Free Rocket Over Ground) applied to a family of relatively heavy unguided battlefield missiles which have been developed and introduced into service in the USSR, the countries of the Warsaw Pact and some others, since the mid-1950s. Six different types of such missile have been identified and are commonly referred to as FROG-1, FROG-2 etc up to FROG-7 with FROG-6 missing. Why FROG-6 is missing is not known.

All six missiles are believed to be or to have been propelled by solid-propellant motors: but there has presumably been at least one change of propellant, since the earliest (but only the earliest) missile was transported on a vehicle fitted with what is generally agreed to have been a heating jacket to prime the propellant.

All but the latest in the series have been seen mounted only on track-laying vehicles: FROG-1 was mounted on a JS III chassis and the next four types were or are mounted on a PT-76 chassis. FROG-7, however, is mounted on a ZIL-135 wheeled chassis and this has a road range more than double that of any of the earlier versions. All six missiles are believed to have a choice of nuclear or high-explosive warheads.

It is believed that FROG-1 and FROG-2 have now been withdrawn from operational formations in the WTO, although it is not unlikely that some are still in use for training purposes. The remaining four types are believed to remain in service at divisional level in the WTO forces; but it seems likely that some of the earlier versions are at least obsolescent. The missiles are widely deployed throughout the WTO, although such nuclear warheads as exist for them must certainly be in USSR custody, and some have certainly been supplied to the Middle East. Others have been supplied to Cuba, North Korea and elsewhere.

FROG-1 was first seen in 1957. It is carried on a tracked vehicle based on the JS III amphibious reconnaissance vehicle, and is covered by a heavy ribbed casing which elevates with the missile to form a launch tube. The container also provides propellant pre-heating for the single-stage motor.

Type: Surface-to-surface artillery missile
Configuration: Cylindrical body with large bulbous warhead with pointed nose. Six tail fins. Tracked transporter has container serving as launcher/erector and with elements for pre-heating missile propellant
Length: 9.5 m
Diameter: 85 cm
Weight: 2 700 kg
Propulsion: Solid
Range: 24-32 km
Guidance: Unguided. Spin-stabilised
Warhead: Optional nuclear or high-explosive

FROG 2 is very similar to FROG 1 but has a different carrier, derived from the PT-76 amphibious reconnaissance tank. There is no casing for the missile on this launcher.

FROG 3 differs from FROG 2 in having two tandem rocket motors as against the latter's single-stage motor and a differently shaped warhead. The same PT-76-based launch/erector is used.

FROG 4 is substantially similar to FROG 3 but most obviously differs in having a warhead of the same diameter as the missile body.

FROG-3 artillery rockets on tracked transporter/erector vehicles

FROG 1 to 3 all have warheads of greater diameter than the remainder of the missile.

FROG 5 is basically very similar to FROG 4, using the same launcher/erector vehicle based on the PT-76 amphibious reconnaissance tank, and having two tandem solid propellant motors. The warhead, though, is noticeably different in shape, being almost perfectly conical and this is assumed to indicate a more advanced payload.

Type: Surface-to-surface artillery missile
Configuration: Cylindrical body with cruciform tail fins. FROG 2 has pointed nose bulbous warhead. FROG 3 warhead is cylindrical with conical nose. FROG 4 has warhead of same diameter as missile body. FROG 5 has conical warhead
Length: 10.2 m
Diameter: 54 cm
Weight: 2 000 kg
(Dimensions are for FROG 4)
Propulsion: Solid. FROG 2: single-stage. FROG 3/4/5: two-stage
Range: Up to 50 km
Guidance: Unguided. Spin-stabilised
Warhead: Optional nuclear or high-explosive

FROG 7 is the latest and most changed of the numerous FROG series of Soviet battlefield support missiles. First shown in public in 1967 this weapon marked a new departure in carrier vehicles, this being a modern, wheeled, erector/launcher designated ZIL-135. The missile itself has a more modern appearance also, reverting to single-stage propulsion and having a warhead of the same diameter as the rest of the missile.

Type: Surface-to-surface atillery missile
Configuration: Cylindrical body with conical nose and cruciform tail fins
Length: 9 m Approx
Diameter: 61 cm Approx
Propulsion: Solid, single-stage
Range: Up to 60 km
Guidance: Unguided. Spin-stabilised
Warhead: Optional nuclear or high-explosive

HONEST JOHN (MGR-1B)

Honest John is a simple surface-to-surface free-flight rocket that has the accuracy of standard artillery with considerably better battlefield mobility. Operational for several years with the US and NATO forces the system has undergone substantial improvement. The unguided rocket is driven by a single-stage solid propellant motor and can carry either a nuclear or a high-explosive warhead.

Designed to fire like conventional artillery in battlefield areas, Honest John is now the oldest missile system still fielded by the US Department of Defense. Studies were begun by US Army Ordnance in 1950, and shortly after that Douglas Aircraft (now McDonnell Douglas) submitted proposals based on the Ordnance specifications and became prime contractor for the system. Firing tests were successfully completed in 1951 at White Sands, New Mexico.

Operation is simple — there are no electrical controls — and normal crew training and standard fire control techniques are employed.

Honest John has been in service with the armed forces of Belgium, Denmark, the Federal German Republic, France, Greece, The Netherlands, South Korea, Turkey, the United Kingdom, and the USA. Production ceased some time ago, and the missile has been replaced in US Army service by Lance. Many other NATO nations are expected to replace Honest John by Lance in due course and the German Federal Republic and the UK have already committed themselves.

Newer FROG-7 artillery rocket with its wheeled transport (left), *and American Honest John training launch* (right)

Type: Surface-to-surface artillery missile
Configuration: Cylindrical body with cruciform delta fins at rear end and slightly bulbous fore section with pointed nose
Length: 7.5 m
Diameter: 76 cm
Span (Max): 1.37 m
Weight: 2 040 kg
Propulsion: Single-stage solid
Range: 7.5-37 km
Guidance: Unguided. Ballistic trajectory
Warhead: Nuclear or high-explosive
Main Contractor: US Army Missile Command

LANCE (MGM-52C) (USA)

Lance is designed to provide general battlefield fire support for an Army Corps. It is replacing Honest John and Sergeant missiles in that role. Both nuclear and a variety of conventional warhead versions have been developed, and other payloads have reached differing stages of advancement. The US Army at present deploys only the nuclear version, but under Army sponsorship Vought has developed the terminally-guided submissile (TGSM) payload for the engagement by indirect fire of tanks or similar targets in the forward area. Six or nine TGSMs would be loaded into each warhead for dispersion over the target area. On release, their individual seekers search for and lock-on to separate targets, guiding the TGSM by means of wrap-around aerodynamic surfaces that erect upon release from the warhead. Initial stabilisation is achieved by means of a para-ballon at the rear of each TGSM. Infra-red homing is the most likely form of guidance but others that are being studied include a millimetre-band radiometric correlator. Yet another payload under consideration is a precision-guided warhead employing Distance Measuring Equipment (DME). Conventional warheads include normal HE and area-cover types, the latter releasing over 800 0.5 kg cluster bomblets.

Tracked and towed versions of the Lance system are produced, the latter being transportable by helicopter. An eight-man crew is normal. In addition to the US Army, Lance is, or will be, deployed by the armed forces of Belgium, West Germany, Israel, Italy, the Netherlands, and the UK. Entry into US service was in 1972.

Type: Surface-to-surface battlefield support missile
Configuration: Cylindrical body with pointed nose and cruciform slender delta tail fins. Self-propelled launch vehicle
Length: 6.1 m
Diameter: 56 cm
Weight: 1 500 kg. Approx
Propulsion: Liquid, pre-packed
Range: 120 km
Guidance: Simplified inertial
Warhead: Nuclear. High explosive warhead under development. Warhead containing multiple battlefield missiles under study (see above)
Main Contractor: Vought Corporation, Michigan Division

TGSM CHARACTERISTICS

Type: Terminally-guided sub-missile. Surface-to-surface, multiple payload for Lance.
Configuration: Cylindrical body, tapered nose carrying seeker head, two sets of four wrap-around fins at tail and near mid-length
Length: 89 cm
Diameter: 15.2 cm
Weight: 15.8 kg
Propulsion: None
Guidance: Not finalised, probably infra-red
Warhead: Armour piercing
Main Contractor: Vought Corporation, Michigan Division

Lance is carried on tracked vehicles (above right) *and in towed configuration* (far right). *Lower picture is of one TGSM that can be carried in warhead*

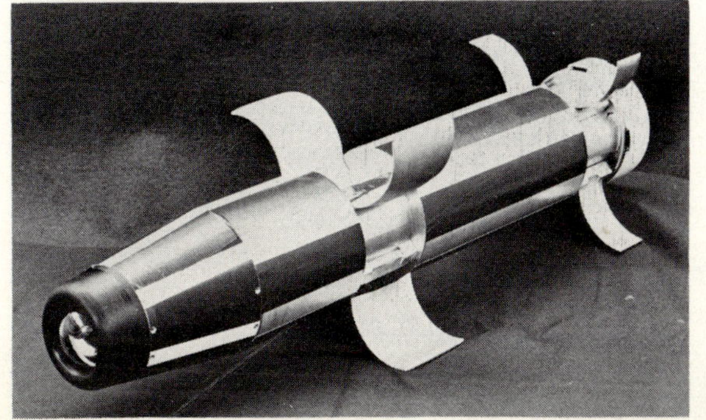

PERSHING (MGM-31A)

Pershing 1A is an improved version of the Pershing 1 ground-to-ground guided weapon system. It has replaced Pershing 1 and is now deployed with the US Army and the Federal German Republic.

The most obvious change is the replacement of the XM-474 tracked vehicles by a set of wheeled vehicles based on the M656 five-ton truck. The system now has an improved erector-launcher which is an articulated truck and trailer combination that carries both the missile and its warhead (previously carried on separate vehicles) capable of both paved road and cross-country travel. The other vehicles are a transporter for the programmer-test and power stations, the firing battery control centre truck, and the radio terminal set vehicle with an inflatable antenna. All the equipment is transported by the CS-130 cargo aircraft. There is no change in the missile itself.

Most recent of the improvements is an Automatic Reference System (ARS) which employs gyro compass techniques enabling the missile to be fired from an unsurveyed site in something like half the time previously required for surveying. Associated with ARS is a sequential launch adapter (SLA) which permits the Pershing commander to count down and launch up to three missiles from a single control station without uncabling and recabling after each missile is launched. European-based Pershing units received ARS/SLA ground equipment in 1976.

A project to provide a system of terminal guidance to increase accuracy and enable a lower-yield warhead to be used (most deployed Pershings are thought to have a 400 kt warhead) is being carried out by Martin Marietta. Started in 1975, this programme is generally known as Pershing II. The proposed guidance technique is Radar Area Correlation Guidance, by which the radar image seen by the incoming warhead is compared with previously obtained reconnaissance imagery to provide correction signals. Six flights of an experimental non-nuclear Pershing 2 warhead were planned for 1977.

Type: Surface-to-surface battlefield support missile
Configuration: Cylindrical body with conical nose section. Three rectangular fins at base of second stage. Three delta control surfaces on first stage
Length: 10.5 m
Diameter: 101 cm
Weight: 4 535 kg
Propulsion: Two-stage solid. Thiokol M-105 first stage; M-106 second stage
Range: 160-740 km
Guidance: Inertial
Warhead: Nuclear. Believed about 400 kilotons
Main Contractor: Martin Marietta Aerospace, Orlando Division

Pershing battery in action

PLUTON

Developed for the French Army, Pluton is a surface-to-surface tactical nuclear missile. The weapon, which has a range capability from 20-120 km, is installed on and fired from the AMX-30 tank chassis, the missile container being used as a launching ramp. Missile and warhead are supplied separately to operational units.

In addition to the AMX-30 launching vehicle the system includes command vehicles containing data processing equipment organised round the IRIS 35M computer of the Plan Calcul Militaire. This is a third-generation general purpose computer capable of operation under severe environmental conditions. If operational circumstances require it, target data can be obtained in real time from an R.20 drone equipped with a Cyclope passive IR reconnaissance system. Missile guidance is by means of a simplified inertial system based on a SFENA semi-strap-down system. Alternative warheads are available: the 25 kiloton yield AN-51 which contains the same MR50 nuclear charge as the AN-52 free-fall bomb carried by French Air Force Mirage III and Jaguar aircraft; or a 15 kiloton nuclear payload. The former is intended for important targets in an enemy's rear, while the smaller yield device would be employed for use nearer to the main battle area.

Initial production contracts were awarded in 1972, and deliveries to the French Armed Forces began early in 1974. It is now operational with four French Armies (3rd, 25th, 60th and 74th) and a fifth (32nd) was in the process of equipping with Pluton during 1977. Each Army has six launchers plus the appropriate support and re-supply units.

Type: Surface-to-surface, tactical
Configuration: Cylindrical body with pointed nose and four tail fins
Length: 7.59 m
Diameter: 65 cm
Span (Max): 1.42 m
Weight: 2 350 kg
Propulsion: Solid
Range: 10-120 km
Guidance: Inertial
Warhead: Nuclear tactical, 25 kt and 15 kt models. Various conventional warheads under study
Main Contractor: Société Nationale Industrielle Aérospatiale

Pluton tactical nuclear missile deployed for action

SCALEBOARD (SS-12) (USSR)

Scaleboard is the NATO name for a short/medium-range ballistic missile which is identified with the SS-12 in the US series. It is a mobile missile normally transported on a MAZ-543 eight-wheeled vehicle similar to that sometimes used for Scud-B. Its warhead has been reported to be in the megaton range. This suggests that it has rather greater destructive power than its approximate equivalent in NATO, the American Pershing missile. Unlike the Scud missiles, which in transit are exposed to view as they rest in their elevating cradles, the Scaleboard missile is enclosed in a ribbed split metal casing which is elevated with the missile into the firing position. Scaleboard appears to be approximately the same length as Scud-B — rather more than 11 m — but larger in diameter. Its range has been estimated in the region of 7-800 km. Nothing is known about its guidance system but it is presumed to be some form of inertial guidance suitable for use with a missile that is elevated from the horizontal to the vertical shortly before firing.

Believed to be fully operational in the USSR, there remains a marked scarcity of reliable information regarding the precise role of this weapon and the numbers and nature of their deployment. In early 1977 a US assessment reported that Scaleboard missiles were still being added to the Soviet inventory. First reported in the West in 1967.

Type: Surface-to-surface, tactical
Configuration: Scaleboard has not been publicly revealed outside the launcher/erector casing which completely envelopes the missile. The following dimensions are based on studies of the container and transporter vehicle (MAZ-543)
Length: 11.25 m
Diameter: 100 cm
Weight: 6 800 kg plus
Propulsion: Presumed storable liquid
Range: 700-800 km. Estimated
Guidance: Presumed inertial
Warhead: Nuclear

*Scud heavy artillery rocket (*right*) exists in several versions and has been supplied beyond the Warsaw Pact states*

The family of heavy artillery rockets known to NATO as Scud, and within US defence circles as SS-1, has been produced in two models, Scud A and Scud B. The more recent Scud B has been noted in two versions one of which is thought to have been an interim stage of development leading to the standard model now widely deployed within the Warsaw Pact forces and those of Iraq, Libya, Egypt and Syria. Scud A is believed to have been withdrawn from service. The possible existence of a longer range Scud C has been reported but not confirmed to date.

Scud is a land-mobile system, single missiles being carried on a vehicle that combines the functions of transporter and erector by means of hydraulically actuated rams which raise the missile to the upright position for launching. Scud A and Scud B (both versions) have been deployed on the tracked JS-III chassis, but from 1965 onward Scud B only has been reported on the MAZ-543 eight-wheeled transporter vehicle.

The standard Scud B missile is capable of carrying either nuclear or conventional warheads but use of the former type is probably restricted to Soviet Union forces. Liquid propulsion is employed, and launch preparation time is quoted by Russian sources as about one hour. Inertial guidance, possibly allied to some form of radio command guidance for the early stages of flight, is the generally assumed method of aiming Scud. Since the fins have every appearance of being rigidly fixed to the body of the missile, control is probably by means of auxiliary vanes positioned within the rocket motor efflux. Whether or not some form of control is provided for the post-burnout phase of flight is not known at present.

The data below refer to the standard Scud B; Scud A is roughly 0.5 metre shorter and had a reported range capability of some 130 km. The range of the provisional Scud C is in the region of 450 km, and both length and diameter of this version are expected to exceed those of Scud B.

Type: Surface-to-surface, artillery missile
Configuration: Cylindrical body with conical nose and cruciform tail surfaces
Length: 11.25 m
Diameter:85 cm
Weight: 6 300 kg. Estimated
Propulsion: Liquid, storable
Range: 160-270 km
Guidance: Inertial
Warhead: HE with nuclear option.

SERGEANT (MGM-29A) (USA)

Sergeant is a field artillery ballistic missile, capable of carrying either a nuclear or a high-explosive warhead, and intended to provide missile support to a corps or field army. A second-generation system, it replaced the Corporal missile and among the improvements offered were inertial guidance and solid-propellant propulsion. These features reduced the amount of ground handling equipment required by the system and resulted in a shorter reaction time than that of the command-guided liquid-fuelled Corporal. Recent improvements in the electronics and procedures for automatic count-down have resulted in a still further reduction in reaction time. Units already in the field have been modified to incorporate these improvements.

The missile is composed of four major sections: warhead, guidance, rocket motor and control surfaces. These sections are assembled just before firing and are transported to the firing position in special sealed containers. Similar missile sections are interchangeable. The guidance system is immune to known countermeasures.

The Sergeant system is in operation with US troops in the United States and overseas. West Germany also has Sergeant. Extensive replacements by Lance are in progress as noted above.

Type: Surface-to-surface battlefield support missile
Configuration: Cylindrical body with slender conical nose section. Cruciform tail surfaces
Length: 10.5 m
Diameter: 79 cm
Span (Max): 1.78 m
Weight: 4 535 kg
Propulsion: Single-stage solid
Range: 45-140 km
Guidance: Inertial
Warhead: Nuclear or high-explosive
Main Contractor: Univac Salt Lake City Company

ANTI-TANK MISSILES

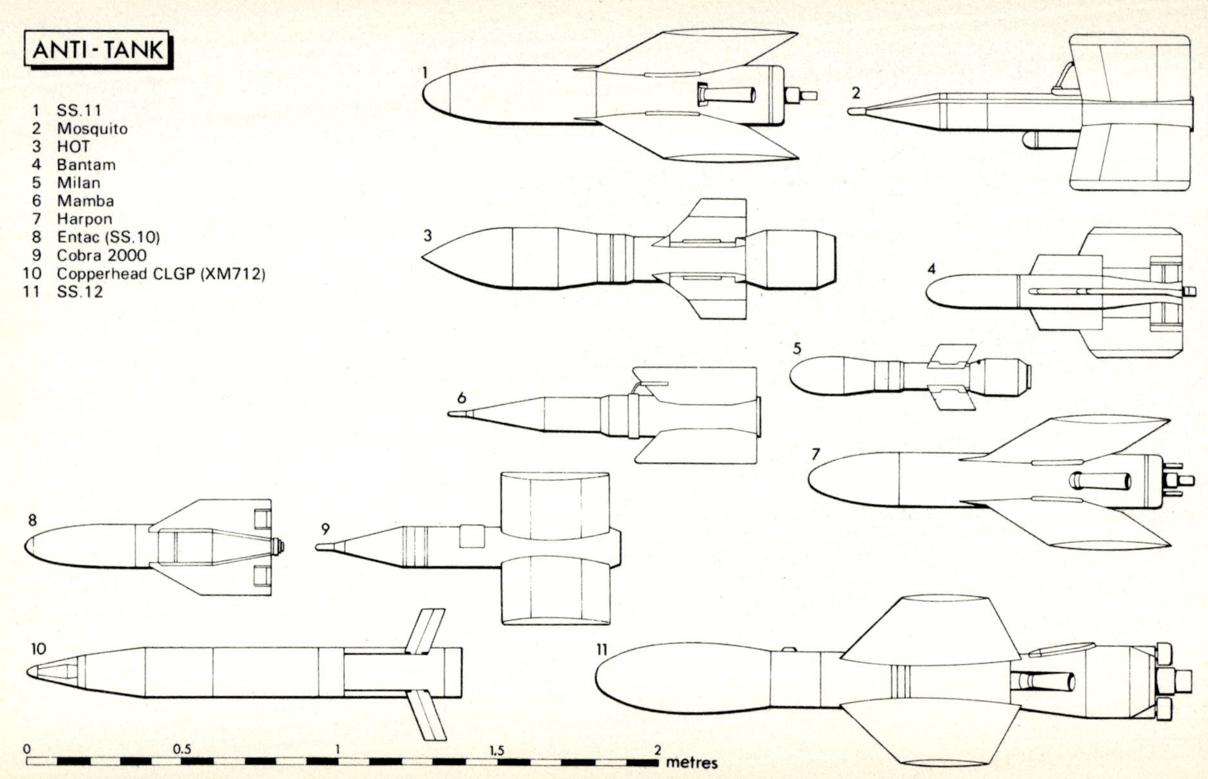
ANTI - TANK
1 SS.11
2 Mosquito
3 HOT
4 Bantam
5 Milan
6 Mamba
7 Harpon
8 Entac (SS.10)
9 Cobra 2000
10 Copperhead CLGP (XM712)
11 SS.12
1
2
3
4
5
6
7
8
9
10
11
0
0.5
1
1.5
2
metres

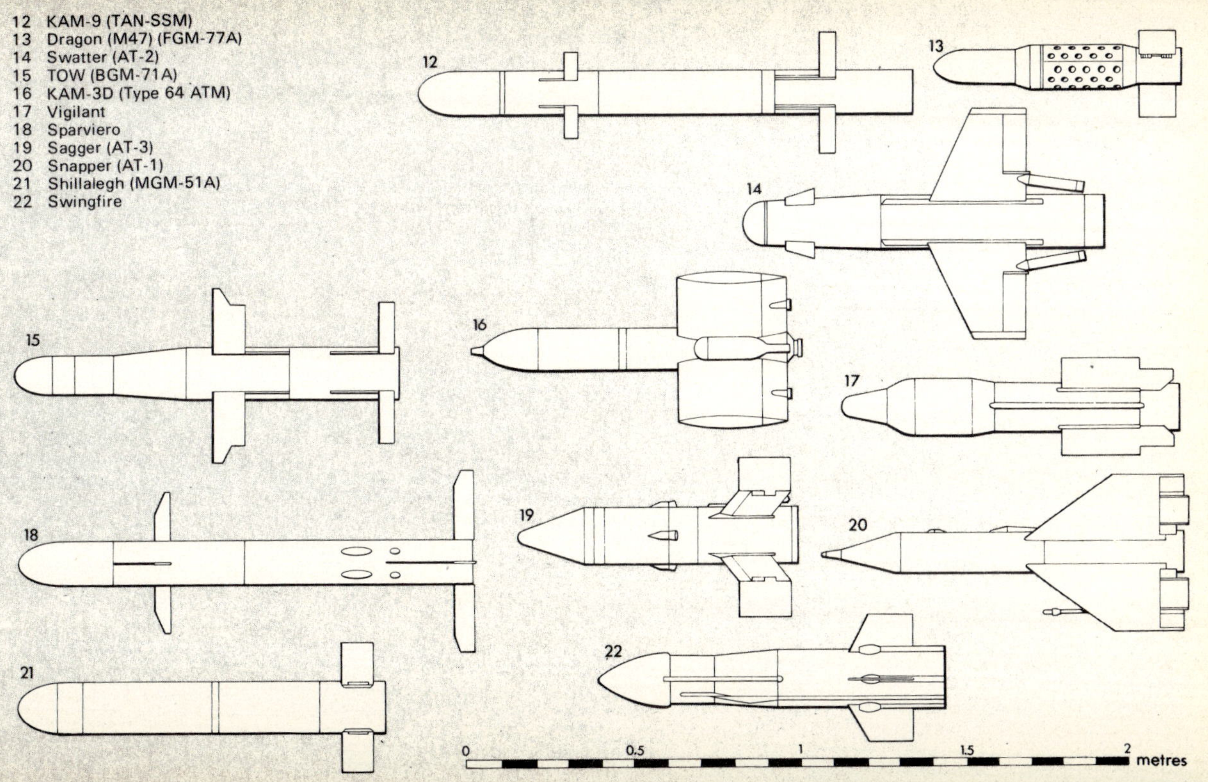
12 KAM-9 (TAN-SSM)
13 Dragon (M47) (FGM-77A)
14 Swatter (AT-2)
15 TOW (BGM-71A)
16 KAM-3D (Type 64 ATM)
17 Vigilant
18 Sparviero
19 Sagger (AT-3)
20 Snapper (AT-1)
21 Shillalegh (MGM-51A)
22 Swingfire
12
13
14
15
16
17
18
19
20
21
22
0
0.5
1
1.5
2
metres

ANTI-TANK MISSILES

Guided Line-of-Sight Systems

Most of the weapons described in this section are small (up to 75kg but generally much less), optically-aimed, guided missiles designed primarily to disable enemy armoured fighting vehicles but usually suitable also for attacking fortified positions. They range in size from systems which are sufficiently small and light to be carried complete by one man to crew served systems which are suitable only for installation in a substantial vehicle. Many can be adapted for use in light aircraft and helicopters, and additional examples of such applications will be found in the section dealing with air-launched weapons. A few systems have also been adapted for naval use.

Reference is frequently made to various generations of such weapon systems. Usage in this respect is not firmly established, but the following rough definitions may be helpful.

First Generation. A weapon system in which the operator observes both missile and target and manually guides the one towards the other. Guidance in such systems is effected by the transmission of electrical guidance signals from the operator's control unit along a pair of thin wires reeled out from the missile as it flies towards the target.

Second Generation. A system in which the operator observes the target and keeps his sight trained on it. The missile is tracked automatically and the displacement between it and the operator's line of sight is used to generate guidance commands which are conveyed to the missile either by means of trailing wires or, as in the American Shillelagh system by some other form of command link.

Third Generation. A system in which, as in a second-generation system, the operator has only to keep his sight trained on the target but in which the deviation of the missile track from the sight-line (defined, for example, by a coded laser beam) is measured by apparatus which is carried by the missile and which generates appropriate guidance commands and applies them directly to the missile guidance system.

Other Guided Systems

One quite different anti-tank system is described in this section: this is the Copperhead CLGP (Cannon-Launched Guided Projectile) system being developed in the USA.

Other Anti-tank / Assault Weapons

In addition to the guided missiles described on the following pages there are a number of shoulder-fired anti-tank rockets in use that do not qualify for inclusion in this book. However the more significant of these unguided missiles are listed briefly.

RL-83 Blindicide Rocket Launcher. 83mm. Belgium
P-27 Anti-Tank Grenade Launcher. 3.75kg grenade. Czechoslovakia
M-55 Anti-Tank Grenade Launcher. 2.75kg grenade. Finland
ACL-STRIM (LRAC Model F1). 2.2kg missile. France
ARPAC Anti-Tank Rocket Launcher. 0.85kg rocket. France
SARPAC Anti-Tank Rocket Launcher. 1.8kg rocket. France
Armbrust-300 (Crossbow). 67mm projectile. W. Gerrmany
PZF-44 Anti-Tank Grenade Launcher. 1.5kg grenade. W. Germany
Folgore Anti-Tank Rocket Launcher. 4.2kg projectile. Italy
M-65 Anti-Tank Rocket Launcher. 2.0kg rocket. Spain
RCL Carl-Gustaf (FFV 550). 3.2kg HEAT round. Sweden
Miniman Recoilless Anti-Tank Weapon. 74mm. Sweden
M-58 Anti-Tank Rocket Launcher. 1.82kg rocket. Switzerland
M-72 Light Anti-Tank Weapon (LAW). 1.25kg projectile. USA
M-20A1 Rocket Launcher. 4.09kg projectile. USA
RPG-7 Anti-Tank Grenade Launcher. 2.5kg grenade. USSR

Bantam (RB 53) is a small wire-guided anti-tank missile, initially designed for infantry use but since developed for vehicle mounting and launching from helicopters and small military aircraft such as the Saab-MFI 17. In its simplest configuration it comprises a missile in its combined transport and launching container, a control unit, and 20 m of cable. The control unit is fitted with a sighting device and a joystick to control the missile's flight path. Among the alternative configurations that have been developed are a six-missile installation for the Puch-Haflinger jeep, a four-missile system for the turret of the Cadillac Gage V-100 AFV, and several airborne versions.

Development was undertaken as a private venture by Bofors in 1956. Bantam is now in series production for the Swedish Army and the Swiss Army, the dates of entry into service being 1963 and 1967, respectively.

Type: Anti-tank, surface-surface, with air-to-surface capability
Configuration: Cylindrical body with rounded nose and rear-mounted cruciform wings
Length: 85 cm
Diameter: 11 cm
Span (Max): 40 cm
Weight: 7.5 kg
Propulsion: Two-stage solid
Range: 250-2 000 m
Guidance: Wire plus gyro stabiliser. Command to line-of-sight
Warhead: Hollow charge armour piercing. 1.9 kg
Main Contractor: AB Bofors

Swedish Bantam anti-tank missile can be surface- and air-launched

COBRA 2000 (W. GERMANY)

This weapon system, for which the full designation is BO 810 Cobra 2000, comprises a small wire-guided anti-tank missile, a control box and cable links. Light in weight and of low cost it is suitable for one-man operation. Unlike many other wire-guided missiles in this category, Cobra is not launched from a container but is simply set on the ground whence it is given a jump start by means of a smaller boost motor. Up to eight missiles can be connected to a single control box and the operator is able to select them in any order for firing. The name Cobra derives from Contraves-Oerlikon-Bölkow Rakete, the two Swiss firms contributing sustainer motors and warheads.

First design studies were carried out in 1957 and the system entered service in 1960, since when more than 150,000 missiles have been delivered to 18 countries. Cobra is in service with the forces of West Germany, Argentina, Brazil, Denmark, Greece, Israel, Italy, Pakistan, Spain and Turkey. Licence manufacture arrangements exist in Brazil, Italy, Pakistan and Turkey.

Type: Anti-tank, surface-to-surface
Configuration: Cylindrical body with conical nose. Cruciform rectangular wings
Length: 95 cm
Diameter: 10 cm
Span (Max): 48 cm
Weight: 10.3 kg
Propulsion: Solid sustainer with non-jettisonable booster
Range: 400-2 000 m
Guidance: Command to line-of-sight wire guidance, with optical gathering and tracking
Warhead: Hollow charge penetration, 2.7 kg, or anti-tank/shrapnel of same weight
Main Contractor: Messerschmitt-Bölkow-Blohm GmbH

Copperhead is the name assigned to the M712 Cannon Launched Guided Projectile (CLGP), a 155mm round equipped with a terminal guidance system and launched from conventional howitzers into a ballistic trajectory. During flight, the target is illuminated by a laser designator operated by a forward observer. The seeker in the nose of the CLGP acquires the semi-active laser signature and uses this signal to operate control surfaces on the projectile to cause it to follow a collision course to the target.

Origin of the system rests in the US Army requirement for a means of bringing indirect fire to bear on tanks and distant armoured targets with a high kill (preferably first round) probability using a shaped charge warhead. Indirect fire demands the use of a forward observer or a forward air controller who calls for fire in the conventional manner and designates the target during the last seconds of the projectile's flight. Several options are provided for this requirement.

The Copperhead CLGP is made up of three sections: guidance, payload, and stabilisation and control. The configuration is aerodynamically controlled by cruciform in-line wings and tail fins that provide roll stabilisation and lateral manoeuvrability sufficient to provide an impact footprint adequate for successfully engaging manoeuvring armoured targets.

The guidance section consists of the seeker and electronic assemblies which are housed within the nose dome and steel housing. The seeker uses folded body-mounted optics with a spin stabilised gimballed mirror (seeker gyro).

The payload section includes an HESH warhead in a steel structure and a fuze module.

The stabilisation and control section includes the flip-out aerodynamic surfaces and associated actuator mechanisms. Deployment of the fins is delayed until after muzzle exit by the effects of launch acceleration. When deployed at their 20° sweep angle, the tail fins are canted to maintain a six to 18rev/sec projectile roll rate, which aids stability and also allows liberal tolerances in vane manufacture.

The M109A1 SP 155mm howitzer and the helicopter transportable 155mm towed howitzer are intended to use the Copperhead round. The former is extensively used within NATO and is able to fire Copperhead in the same way as current standard ammunition.

The Copperhead round is fired in an inactive state, thus simplifying the precautions needed to protect the guidance system from acceleration shocks. An acceleration-sensitive battery, initiated by firing the projectile, supplies power for guidance and control, as well as initiating and operating a timer which controls the internal operating sequence.

Associated with timing is the selection of the trajectory mode for a ballistic trajectory or a fly under fly out (FUFO) trajectory. The latter is intended for ranges of over 8km and for bad weather where the projectiles travel at a lower altitude to provide a longer time beneath cloud ceiling for acquisition and guidance. The ballistic trajectory is used for shorter range shots and/or good weather conditions.

Coding of the laser signature which will be accepted by the seeker is also set in at the same time as the above selections are made. This coding will match that being used by the forward operator of the laser designator.

The 155mm CLGP was conceived by engineering staff of the US Army's Rodman Laboratories in 1970, and feasibility studies were conducted during 1971 by the Army Missile Command, Picatinny Arsenal, Rodman Laboratories, and the US Naval Weapons Laboratory at Dahlgren under the direction of the Army Project Manager, Canon Artillery Weapons Systems. Two Advanced

Development contracts were awarded in February 1972, to Martin Marietta and Texas Instruments, and on completion each contractor was asked to submit proposals for Engineering Development of a 155mm CLGP.

The Martin Marietta design was selected following tests involving 12 rounds from each competitor. Two of the Martin projectiles were retained for demonstration with airborne target designation systems. One of these hit a tank marked by a Mini-RPV on 3 October 1975, and on 26 February 1976 a moving tank was hit at a range of 8km in darkness with target designation from a Cobra helicopter hovering some 3 km from the target. Sums of $36.1m and $54.1m were to be spent on research, development and engineering during 1977 and 1978.

Type: Anti-tank, surface-to-surface
Weight: 63.5 kg
Length: 137.2 cm
Warhead weight: 22.5 kg
Explosive weight: 6.4 kg
Maximum range: 20 km
Minimum range: 3 km
Main Contractor: Martin Marietta Aerospace, Orlando.

*M712, Copperhead, cannon-launched guided missile (*right*), and Dragon anti-tank missile ready to fire (*far right*)*

DRAGON (M47) (FGM-77A)

At one stage known as MAW (Medium Anti-tank assault Weapon) Dragon has been developed for the US Army and Marine Corps as a weapon light enough to be carried and fired by one man yet powerful enough to destroy most armour and other hard targets encountered on the battlefield. Three items comprise the system: a tracker, a recoiless launcher, and the missile. The tracker includes a telescope for the gunner to sight the target, a sensor and an electronics package. The missile features a unique propulsion system consisting of 30 pairs of small rocket motors mounted in rows around the missile body.

Command-to-line-of sight guidance is employed with an infra-red sensor determining the relative position of the missile. This results in automatic command signals being transmitted to the missile's rocket motors via trailing wires.

A $35 million contract for design, development and testing was awarded to McDonnell Douglas in late 1966 and a $30 million award followed in 1967 for production engineering. Production began in 1972 and in September of that year Raytheon was chosen as a second source, with Kollsman as second source contractor for the sight. In addition to the US Army and Marine Corps, orders have been placed by Iran, Israel, Saudi Arabia, and Switzerland (sample batch only). Planned procurement for fiscal years 1976/7/8 is 34,443, 16,080, and 20,671 rounds, respectively.

Type: Anti-tank, surface-to-surface, man portable
Configuration: Cylindrical body with rear section of larger diameter. Three curved fold-out fins at rear. Tube-launched
Length: 74 cm
Span (Max): 33 cm
Weight: 12.25 kg
Propulsion: Multiple solid rocket motors disposed in pairs in centre section of body
Range: 1 000 m
Guidance: Wire. Command to line-of-sight
Warhead: High-explosive shaped-charge
Main Contractors: McDonnell Douglas Astronautics Company, Raytheon Company, Kollsman Instrument Company

ENTAC (SS.10) (FRANCE)

Entac (Engin Téléguidé Anti-Char) is a first-generation, roll-stabilised, wire-guided anti-tank weapon which, although no longer in production, is in widespread service. Since its adoption by the French Army in 1957 it has been delivered to Australia, Belgium, Canada, India, Indonesia, Iran, Morocco, Norway, South Africa, Switzerland and the USA. In the last of these countries it is designated (MGM-32A). The manufacturers report a total production of 119,409.

It was developed by the Direction Technique des Armements Terrestres (DTAT) as a replacement for the earlier SS.10 for use by infantry, mechanised infantry and parachute units. A single operator can control up to 10 remotely sited missiles, the maximum distance between operator and launcher being 110 metres. The standard system employs a four-round launcher but helicopter, jeep and other vehicular mountings have been developed. Milan is replacing Entac in a number of countries.

Type: Anti-tank, surface-to-surface
Configuration: Short cylindrical body tapered at both ends. Rear-mounted cruciform wings
Length: 82 cm
Diameter: 15 cm
Span (Max): 37.5 cm
Weight: 12 kg
Propulsion: Solid
Range: 2 km maximum
Guidance: Wire-guided, roll stabilised. Control by wing spoilers
Warhead: High-explosive, 4.1 kg
Main Contractor: Société Nationale Industrielle Aérospatiale

HARPON

Harpon is a wire-guided battlefield weapon intended for use on land vehicles, ships, fixed-wing aircraft and helicopters. It has the same characteristics, performance and alternative warheads as the SS.11 B1 weapon but has a greatly improved and automatic form of guidance. This is the improved TCA (Télécommande Automatique) system used by a number of other French missiles such as the AS.30. While the operator keeps his optical sight steady on the target, a separate infra-red tracker follows the missile with the aid of flares attached to it. The signals provided by the respective target and missile tracking systems are used to derive command signals to keep the missile on the line-of-sight to the target.

Harpon is generally deployed on light armoured vehicles such as the Panhard armoured car, or more recently, the AMX-13 tank. In the former instance a four-round turret launcher is provided while a rack attached to the main gun turret carries four Harpon missiles on the AMX-13. A similar arrangement is employed for Harpon on Saudi Arabian AMX-30 tanks. French, West German and Saudi Arabian forces operate Harpon.

Type: Anti-tank, surface-to-surface
Configuration: Cylindrical body with cruciform wings
Length: 1.2 m
Diameter: 16.5 cm
Span (Max): 50 cm
Weight: 30 kg
Propulsion: Two-stage solid
Range: 350-3 000 m
Guidance: Wire, with automatic command to line of sight using stabilised optical sight
Warhead: Shaped charge. High explosive
Main Contractor: Société Nationale Industrielle Aérospatiale

Harpon anti-tank missile

HOT (INTERNATIONAL)

HOT (Haut subsonique Optiquement téléguidé tiré d'un Tube) is a heavy anti-tank weapon developed by Aérospatiale and Messerschmitt-Bölkow-Blohm. With low-speed spin-stabilisation like Milan, it is a tube-launched, wire-guided missile of larger size, with a higher performance than Milan but has the same general principles of operation. HOT is planned as a replacement for the SS.11 missile, and its mission profile corresponds to a NATO requirement for a missile to operate primarily from armoured or unarmoured vehicles and helicopters.

Like Milan, the missile has tail fins which fold down against the body when it is in its launch tube and open out to spin stabilise it in flight. Because of its comparitively high speed (950 km/h) the time of flight to a target is only about one half of that for the SS.11. Guidance is by means of the TCA type of optical/infra-red system developed for the Harpon missile, with optional manual guidance.

A number of launcher installations have been developed: the manual control two-round model for the M.113 vehicle; the UTM 800 electric four-round version being produced for the Panhard M3 AFV; the TH20 electric four-round system for the AMX-10 APC; the K3E automatic loader/launcher for West German Raketen Jagdpanzer; and heliborne launchers for Alouette III, SA 341 and SA 342 Gazelle, and BO 105 helicopters. HOT is in production for French and German Forces, and contracts have been placed by other countries.

Type: Anti-tank surface-to-surface or air-to-surface
Configuration: Circular cross-section body with fore and aft sections of larger diameter. Pointed nose. Cruciform folding wings. Tube launched
Length: 1.28 m
Diameter: 14 cm
Span (Max): 31 cm
Weight: 21 kg
Propulsion: Two-stage solid
Range: 75 m-4 km
Guidance: Command to line-of-sight by wires. Optical aiming with automatic infra-red missile tracking. Manual tracking optional
Warhead: Hollow charge, high penetration
Main Contractors: Messerschmitt-Bölkow-Blohm GmbH, and Société Nationale Industrielle Aérospatiale (Euromissile)

KAM-3D (Type 64 ATM)

The KAM-3D is standard equipment for the Japanese Ground Self-Defence Force, and is intended for anti-tank operations. It can be fired singly or in multiple units by infantry and is also carried by Jeeps and helicopters. When launched by infantry it is fired from a tubular metal launcher/carrying rack at an elevation angle of 15 degrees. The operator controls KAM-3D by means of an optical tracker and a hand-held controller. A two-man operational team is required in the field.

Following a design study carried out in 1956, development was started in 1957. The KAM-3 was adopted as standard in 1964 after several hundred test rounds had been fired. The official designation given then was Type 64 ATM.

Type: Anti-tank, surface-to-surface or air-to-surface
Configuration: Cylindrical body with cruciform rectangular wing
Length: 1 m
Diameter: 12 cm
Span (Max): 60 cm
Weight: 15.7 kg
Propulsion: Two-stage solid
Range: 350-1 800 m
Guidance: Wire plus gyro stabiliser. Command to line-of-sight with optical tracking
Warhead: High-explosive
Main Contractor: Kawasaki Heavy Industries Ltd

HOT anti-tank missile has been mounted on numerous land vehicles and helicopters. Four-tube installation on left is on Panhard M3 APC

KAM-9 (TAN-SSM) (JAPAN)

KAM-9 is an extended range, higher performance version of the KAM-3 anti-tank missile (which see), and which can be used against armoured vehicles on both land and water. Launching is from a tubular container which is also used for storage and transport. A separate solid motor ejects the missile from the launcher and carries it to a safe distance before the main flight motor is ignited. Prior to launch the container is placed on a launch and tracking unit which consists of the firing mechanism, sight, and missile checkout device.

Development was instituted by the Technical R&D Institute, Japan Defence Agency, and the first design study was started in April 1964. Quantity production is expected to begin shortly, although pre-series production is reported to have taken place in 1974.

Type: Anti-tank, surface-to-surface
Length: 1.5 m
Diameter: 15 cm
Span (Max): 33 cm
Propulsion: Solid booster. Two-stage solid sustainer
Guidance: Wire. Command to line-of-sight with optical aiming and automatic (probably infra-red) tracking
Warhead: Armour piercing
Main Contractor: Kawasaki Heavy Industries Ltd

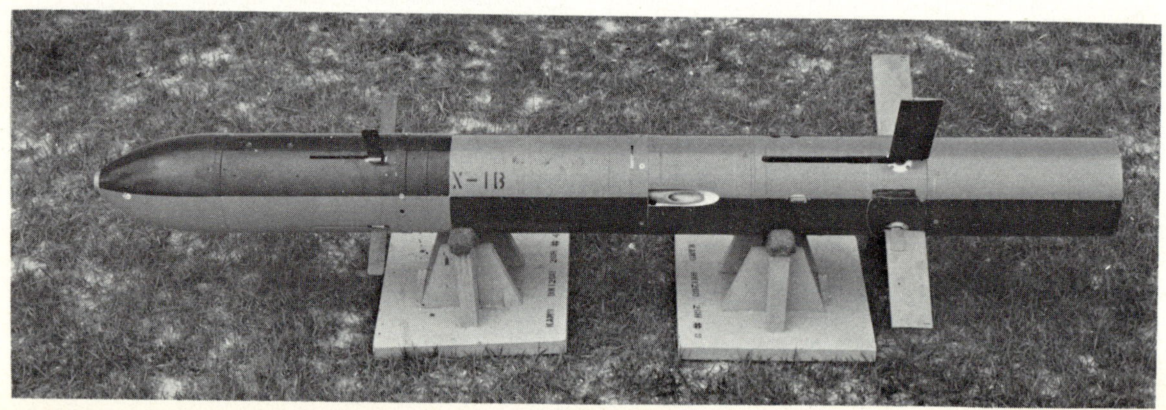

MAMBA

The Mamba anti-tank missile was first announced by MBB in late 1972. Of the same general type as Cobra it incorporates two major improvements in the missile among numerous detail changes, and MBB has developed a completely new controller for the system. The new missile, however, will accept Cobra warheads and can be used with existing Cobra controllers. Similarly, Cobra vehicle launchers can be modified for use with Mamba. A major difference between the two missiles is that the newer Mamba has a one-cell dual-purpose oblique engine which not only provides the jump start facility but provides weight compensating thrust during cruise, whereas the Cobra has separate boost and cruise motors. Mamba is also faster with a speed of 140 m/sec. The controller is capable of operating up to eleven missiles.

Type: Anti-tank, surface-to-surface
Configuration: Cylindrical body with conical nose and cruciform rectangular fins
Length: 95 cm
Diameter: 12 cm
Span (Max): 40 cm
Weight: 11.2 kg
Propulsion: Single-cell, dual-purpose, oblique thrust, solid
Range: 300-2 000 m
Guidance: Wire guidance. Command to line-of-sight with optical tracking
Warhead: Hollow charge armour piercing 2.7 kg, or anti-tank/shrapnel of same weight
Main Contractors: Messerschmitt-Bölkow-Blohm GmbH, Contraves, Oerlikon

Japanese KAM-9 (left) and West German Mamba (right)

MILAN

(INTERNATIONAL)

Milan (Missile d'Infantri Léger ANti-char) is a second generation wire-guided anti-tank weapon which is normally deployed as a man-portable, tripod launcher system. Vehicle-mounted systems have also been developed. The tube launcher forms a transport container also, which is attached to a launch and control unit for firing. This unit incorporates the sight and tracker elements of the French TCA optical/infra-red guidance system used for other missiles such as the SS.11 (which see), HOT, and the air-to-surface AS.30.

Agreement between the French and German partners was reached in 1962 and development began in 1963. Technical evaluation and firing trials for the forces of these two nations were completed in 1971. User trials followed.

First batches of 22,000 missiles and 800 launch units have been ordered for both French and West German armies. Further orders are expected. Large-scale production has been planned on the assumption of a total market of 200,000 missiles and 5,000 launch units. Current production is running at 2,000 missiles and 130 launch units per month. Licence production for the British Army is to be undertaken by BAC and several other NATO countries are adopting Milan.

Type: Anti-tank, surface-to-surface
Configuration: Circular cross-section body with fore and aft ends of larger diameter. Rounded nose. Cruciform folding wings. Tube launched
Length: 75 cm
Diameter: 11.6 cm
Span (Max): 26.6 cm
Weight: 6.5 kg
Propulsion: Solid. Booster ejector in launcher tube
Range: 25-2 000 m
Guidance: Wire Automatic command to line-of-sight by optical tracker/sight
Warhead: Hollow charge
Main Contractors: Messerschmitt-Bölkow-Blohm GmbH and Société Nationale Industrielle Aérospatiale (Euromissile)

(ITALY)

MOSQUITO

Mosquito is a one-man anti-tank weapon of conventional design consisting of a cylindrical glass-fibre body and folding cruciform wings of sandwich construction. For storage and transport, a single Mosquito is packed into a container, with its wings in place and only the warhead detached. The complete package with missile inside weighs 22 kg. Six Mosquitos can be transported, ready for firing, on a Puch-Haflinger light cross-country vehicle, and this weapon has also been mounted on Agusta-Bell 47 helicopters.

Initial design was carried out by the Swiss Contraves concern, but subsequent manufacture was at the company's Italian plant. In service with the Italian Army but production ended.

Type: Anti-tank, surface-to-surface
Configuration: Cylindrical body with conical nose. Cruciform folding wings
Length: 1.11 m
Diameter: 12 cm
Span (Max): 60 cm
Weight: 14.1 kg
Propulsion: Two-stage solid
Range: 360-2 300 m
Guidance: Wire-guided. Manual control, optical tracking
Warhead: Hollow charge or fragmentation. 4 kg
Main Contractor: Contraves Italiana, SpA

Milan set up and ready to fire (left) and the remotely fired Mosquito (right)

SAGGER (AT-3) (USSR)

Sagger is the NATO code name for a small wire-guided anti-tank missile, also known by the US code AT-3, and first seen in the VE-Day anniversary parade in Moscow in 1965. More compact than either the AT-1 (Snapper) or AT-2 (Swatter) missiles, but apparently carrying an equally powerful warhead, it is deployed in various vehicle mountings including a 2x3 mounting on the BRDM-1 amphibious reconnaissance car and a single mount on the BMP-1 armoured personnel carrier. A more recent installation is on the new BMD light tank, and it can also be deployed on the Hind-A helicopter. Other vehicle installations are the BRDM-2/BTR40PB and the Czechoslovakian SKOT OT-64 Model 5 armoured personnel carrier. There is also a single manpack mounting that enables it to be fired from the ground.

In the BRDM vehicle the two clusters of three missiles are mounted retractably and are shielded when in the firing position by the cover plate that protects them when in the retracted position. Like the two earlier missiles, the AT-3 is powered by a solid-propellant rocket motor.

Sagger is in widespread service within the Warsaw Pact armies and in Egypt, Syria, Vietnam and Yugoslavia.

Type: Anti-tank, surface-to-surface
Configuration: Cylindrical body with conical nose. Cruciform wings set near rear end
Length: 86 cm
Diameter: 12 cm
Span (Max): 47 cm (provisional)
Weight: 11.3 kg
Propulsion: Solid
Range: 500-3 000 m
Guidance: Wire Command
Warhead: High-explosive Anti-tank

Russian amphibious armoured infantry combat vehicle (left) *is armed with a 7.62mm machine gun, 73mm cannon and Sagger anti-tank missile. American tank* (right) *fires Shillelagh guided missile from 152mm gun*

SHILLELAGH (MGM-51A)

Shillelagh is a lightweight close-support army guided weapon system intended primarily for use in a ground-to-ground role but suitable also for air-to-surface launching from helicopters. The missile is fired from a 152mm dual-purpose gun and missile launcher which also can fire conventional ammunition. Shillelagh is fired by a gunner, who guides the missile to its target using an infra-red command guidance system. He merely points the cross hairs at the target and follows the target during the missile flight to the point of impact. A missile tracker associated with the gunner's telescope measures the deviation of the missile's flight path from the line of sight, and the resultant signals are converted into commands that are transmitted by the infra-red transmitter to the missile receiver. Here the commands are translated into actuation signals for the jet reaction flight controls. The missile is fitted with flip-out fins which open when it leaves the launcher.

Shillelagh research and development began in 1959 and limited production started in 1964. Deliveries to the US Army commenced in 1966 with deployment the following year. Since then it has been made in larger numbers than any previous American guided missile. It is the main armament of the M551 Sheridan armoured reconnaissance vehicle and has been adapted to the M60A2 main battle tank. During 1975 test firings of an experimental laser-guided version were carried out under a DoD Advanced Research Projects Agency programme. The laser replaces the gunner's optical line-of-sight tracking of the target by a laser marking technique. In April 1976 modifications of a Sheridan and an M20A2 tank took place for tests of the Shillelagh LBR (Laser Beam Rider).

Type: Surface-to-surface close support weapon
Configuration: Cylindrical body with hemispherical nose. Four small flip-out fins at rear. Combination gun launcher
Length: 1.14 m
Diameter: 15.2 cm
Weight: 27 kg
Propulsion: Cannon launch. Single-stage solid sustainer
Guidance: Infra-red command to line-of-sight
Warhead: Shaped charge
Main Contractor: Aeronutronic Ford Corporation

SNAPPER (AT-1) (USSR)

Snapper is the NATO Code name for the wire-guided anti-tank missile believed to be known to the Russians as Shmell (Bumblebee) or 3M6 and known also by the US code AT-1. It is a vehicle-borne system, now usually deployed on the BRDM armoured amphibious vehicle, and is similar in general configuration to such missiles as the MBB Cobra and Contraves-Oerlikon Mosquito.

The missile is launched from a guide rail, a triple mounting being standard in the BRDM vehicle, whereas quadruple mountings are used on the GAZ-69 light cross-country vehicle on which the missile was first deployed. The triple mounting is retractable, the weapons being transported under cover plates which open up for firing. There is also a ground launcher for infantry use.

Provided with periscope binoculars embodying an illuminated variable-brightness reticle with which to sight the target, the operator uses a joystick to control the missile, keeping it on the line of sight to the target with the aid of tracking flares on two of the wings. The missile can be fired and guided by an operator stationed anywhere up to 50 m from the launcher.

Although previously thought to have been withdrawn from service within the Warsaw Pact countries, Snapper has recently been reported as deployed by Afghanistan, Bulgaria, Czechoslovakia, Egypt, East Germany, Hungary, Mongolia, Poland, Romania, Syria, USSR, and Yugoslavia. Nevertheless, many western observers consider this missile to be obsolete, and in the process of being replaced by the AT-3 Sagger.

Type: Anti-tank, surface-to-surface
Configuration: Cylindrical body with conical nose. Cruciform delta wings set at rear end. Ramp launched
Length: 1.13 m
Diameter: 14 cm
Span (Max): 75 cm
Weight: 22.25 kg
Propulsion: Solid
Range: 2 330 m Max
Guidance: Wire. Command to line-of-sight. Optical tracking
Warhead: Hollow charge armour piercing, 5.25 kg. Contact fuze

Soviet Snapper (left) and Italian Sparveiro (opposite)

SPARVIERO

Sparviero (Hawk) is a third-generation anti-tank missile being developed for service in the 1980s. It is claimed to be the first missile of this generation to dispense with wires for the transmission of guidance command signals. An infra-red guidance system developed by Officine Galileo will be used. All-up weight of the system, which consists of a tripod-mounted launcher with sight and guidance package, is 69 kg, while the missile accounts for 16.5 kg of this. Replacement rounds in their transport containers weigh 19 kg. Initial launch velocity of the missile is 60 m/s, reaching a maximum of 290 m/s, with a total flight time of 10.5 seconds.

The principal version being developed is for infantry use, but other versions for armoured personnel carriers, other vehicles, and helicopters are also in progress.

Type: Anti-tank, surface-to-surface and air-to-surface
Configuration: Cylindrical body with rounded nose, slender flip-out cruciform vanes at rear and near nose. Tube-launched
Length: 1.38 m
Diameter: 13 cm
Span (Max): 53 cm
Weight: 16.5 kg
Propulsion: Solid
Range: 75-3 000 m
Guidance: Command-to-line of sight. Infra-red
Warhead: 4 kg hollow charge high explosive
Main Contractor: Breda Meccanica Bresciana, Officine Galileo

SS.11

(FRANCE)

The SS.11 wire-guided battlefield missile has been applied to varied roles and adapted for use with most types of launcher platform since its introduction and it is in use for vehicle, ship and aircraft applications. The current version is the SS.11 B1 model which was introduced in 1962 and uses solid-state electronics in the firing equipment. There is a range of different warheads to suit varying operational requirements. These include the Type 140AC anti-tank warhead capable of perforating 60 cm of armour plate, Type 140 AP02 explosive semi-perforating anti-personnel warhead (2.6 kg of explosive) which will penetrate an armoured steel plate 1 cm thick at a range of 3,000 m and explode about 2.1 m behind the point of impact, and the Type 140 AP59 high-fragmentation anti-personnel type with contact fuze. In addition to France, the SS.11 has been adopted by about 28 countries and 163,000 have been produced. The air-to-surface model is designated AS.11, the US code is AGM-22A, and the original manufacturer's number was Nord 5210. It was superseded by the Harpon in 1967-8. Licence manufacture has been undertaken in West Germany and in India.

Type: Anti-tank, surface-to-surface and air-to-surface tactical missile
Configuration: Cylindrical body with cruciform wings which provide lift and autorotation
Length: 1.2 m
Diameter: 16.5 cm
Span (Max): 50 cm
Weight: 30 kg
Propulsion: Two-stage solid
Range: 350-3 000 m
Guidance: Wire guided
Warhead: Various, see text above
Main Contractor: Société Nationale Industrielle Aérospatiale

Scout helicopter is one of many launchers for SS.11

SS.12

Of similar design to the SS.11, the SS.12 has found two main applications as the SS.12M and AS.12 versions. In the former a special guidance wire for use over water is employed to suit the system to naval applications and a gyro-stabilised sight is used. The AS.12 model has remote-control wires similar to those of the SS.11, wound on twin spools. An APX260 or APX334 gyro-stabilised sight is used. Among the types capable of launching the AS.12 are the Atlantic and Nimrod maritime aircraft and Wasp, Wessex, and Lynx helicopters. In the land-based role the SS.12 has the power for effective use against tanks and other 'hard' battlefield targets. The marine version, SS.12M, was first demonstrated in 1966 and has since been deployed on the vessels of 17 navies, including three Vosper-Thornycroft ships of the Libyan Navy which carry four-round launchers capable of operating either SS.12M or the smaller SS.11M. By late 1976 production of the SS.12M had reached 1,800; and of the AS.12, 4,200.

Of the numerous countries equipped with one or more versions of this missile, the following have been revealed: France, Germany, Italy, Libya, Netherlands, UK.

Type: Surface-to-surface (SS.12) or air-to-surface (AS.12) tactical missile
Configuration: Cylindrical body with slightly bulbous nose. Cruciform wings
Length: 1.87 m
Diameter: 21 cm
Span (Max): 65 cm
Weight: 75 kg
Propulsion: Two-stage solid
Range: 8 km
Guidance: Wire guided, control by varying thrust of two sustainer jets on sides of missile
Warhead: 30 kg armour piercing, shaped charge, or fragmentation
Main Contractor: Société Nationale Industrielle Aérospatiale

SWATTER (AT-2) (USSR)

Swatter is the NATO code name for the anti-tank missile which is known as the AT-2 in the US code and which, like the AT-1 (Snapper), is carried on the BRDM amoured amphibious vehicle.

Of similar size to the AT-1, but slightly heavier (26.5 kg), it is a more advanced missile and has a different configuration. Control is by elevons mounted on the training edges of the rear-mounted cruciform wings.

The standard mount on the BRDM vehicle carries four missiles mounted on guides. Radio command-to-line-of-sight guidance is employed, the command link having two frequencies as protection against electronic counter-measures. There is also the possibility of separate terminal homing (probably infra-red) being provided, and this with radio guidance suits the Swatter for airborne roles such as part of the armament of the Mi-24 Hind-A attack helicopter. It has been deployed with Warsaw Pact forces and with those of Egypt and Syria.

Type: Anti-tank, surface-to-surface and air-to-surface
Configuration: Cylindrical body with rounded nose. Rear set cruciform wings. Two small foreplanes. Ramp launched
Length: 1.12 m Provisional
Diameter: 15 cm Provisional
Span (Max): 66 cm Provisional
Weight: 26.5 kg
Propulsion: Solid
Range: 600-2 500 m
Guidance: Radio. Command to line-of-sight, possibly with terminal homing
Warhead: High-explosive armour piercing

SWINGFIRE

Swingfire is a wire-guided anti-tank missile developed from an original design by Fairey Engineering started in 1958. Subsequently taken over by BAC, the programme continued and entry into service with the British Army was in 1969. The name derives from the system's 90-degree engagement arc from a fixed launch position. Armoured vehicle, infantry, and helicopter variants of the system have been developed, and a feature of the system is the availability of a separation sight which allows the operator to control one or more missiles from a remote position. This can be used for vehicle- or pallet-mounted launcher, while AFVs generally incorporate an integral periscopic sight. Vehicles that have been fitted with Swingfire include the FV712 Ferret Mk 5 scout car (four rounds), FV438 APC (two rounds, plus 12 stowed), Striker combat reconnaissance vehicle, and Land Rover using a six-round pallet-type mounting. More recently a four-round system based on the Argocat vehicle has been developed, this being particularly suited to desert operations. Another version consists of a single-round Swingfire container to which is attached a pair of wheels for towing by an infantryman in the fashion of a golf trolley, hence the popular name, Golfswing. A helicopter version designed for rapid fitting to aircraft such as the Lynx is known as Hawkswing. The infantry version of the system has been named Beeswing. In addition to the British Army, Swingfire systems have been ordered by Belgium and Egypt.

Type: Anti-tank, surface-to-surface. Air-launched version also
Configuration: Cylindrical body with flip-out wings at rear end. Container launched
Length: 1.07 m
Diameter: 17 cm
Span (Max): 37 cm
Propulsion: Solid
Range: Less than 150 m to 4 000 m
Guidance: Wire, command to line-of-sight. Optical sighting
Warhead: Hollow charge armour piercing
Main Contractor: British Aircraft Corporation

*BRDM armoured car (*left*) carries four Swatter anti-tank missiles. Swingfire can be deployed in Golfswing infantry configuration (*right*) or on vehicles or helicopters*

TOW (BGM-71A) (USA)

TOW (Tube-launched Optically-tracked Wire-guided) is a heavy assault anti-tank weapon which has been developed for a variety of land-based applications and for helicopter operations. For infantry use, the system comprises five elements, none of which weighs more than 24 kg although the complete launcher weighs 78 kg when assembled and ready to fire.

TOW can also be installed in most of the available wheeled or tracked vehicles capable of cross-country travel, and the US Army is using it on APCs, jeeps and UH-1B and AH-1G HueyCobra helicopters. An agreement between the US and Italian Governments provides for TOW to be fitted to Agusta A109 helicopters of the Italian Army Light Aviation force. The missile will also arm the US Army Advanced Attack Helicopter. The US Marine Corps has tested it on the "mechanical mule".

TOW is in full production and other users, besides the USA, include Canada, Denmark, W Germany, Greece, Iran, Israel, Italy, Jordan, Kuwait, Luxembourg, Netherlands, Norway, Oman and Turkey. By early 1975 about 150,000 TOW missiles had been ordered. The annual totals for 1976/7/8 are projected as 26,926, 13,051, and 14,866, respectively. Research and development is in progress on programmes to provide extended range and night firing capabilities.

Type: Anti-tank, surface-to-surface and air-to-surface
Configuration: Cylindrical body with rounded nose section of slightly smaller diameter. Cruciform flip-out wings at mid-length, cruciform flip-out tail fins indexed at 45 degrees to wings. Tube-launched
Length: 1.17 m
Diameter: 15.2 cm
Weight: 18 kg
Propulsion: Two-stage solid
Range: 65-3 000 m
Guidance: Wire. Command to line-of-sight automatic optical tracking
Warhead: Shaped charge, armour piercing
Main Contractor: Hughes Aircraft Company

TOW (Tube-launched, Optically-tracked, Wire-guided) anti-tank missile (left) and Vigilant mounted on a scout car (right)

VIGILANT

Vigilant is a man-portable wire-guided anti-tank system for use by infantry, armoured regiments or paratroops. In its simplest configuration the system comprises a launcher box containing one missile, a sight/controller, a pocket battery and a length of interconnecting cable. Optional additions are a selector box to enable one operator to control up to six missiles in turn, a remotely traversable launcher containing two missiles, and a selector box that enables an operator to control three such launchers. Other system variations include mountings for scout cars and other light military vehicles.

Vigilant was developed by Vickers-Armstrongs (Aircraft) Ltd as a private venture from a design specification issued in 1956. Test firings were made in 1957-8 and the production weapon was tested in 1960. It has since been widely sold and went into service in 1963 with British, Finnish and Kuwaiti forces. Other users are Saudi Arabia (1964), Libya (1968), and Abu Dhabi (1971).

Type: Anti-tank, surface-to-surface
Configuration: Cylindrical body comprising several sections of different diameters. Conical nose. Cruciform rectangular wings at rear. Container-launched
Length: 1.07 m
Diameter: 11 cm
Span (Max): 28 cm
Weight: 14 kg
Propulsion: Two-stage solid
Range: 230-1 375 m
Guidance: Wire. Manual or automatic command to line-of-sight. Optical sighting
Warhead: Hollow charge armour piercing. More than 6 kg
Main Contractor: British Aircraft Corporation

AIR DEFENCE MISSILES

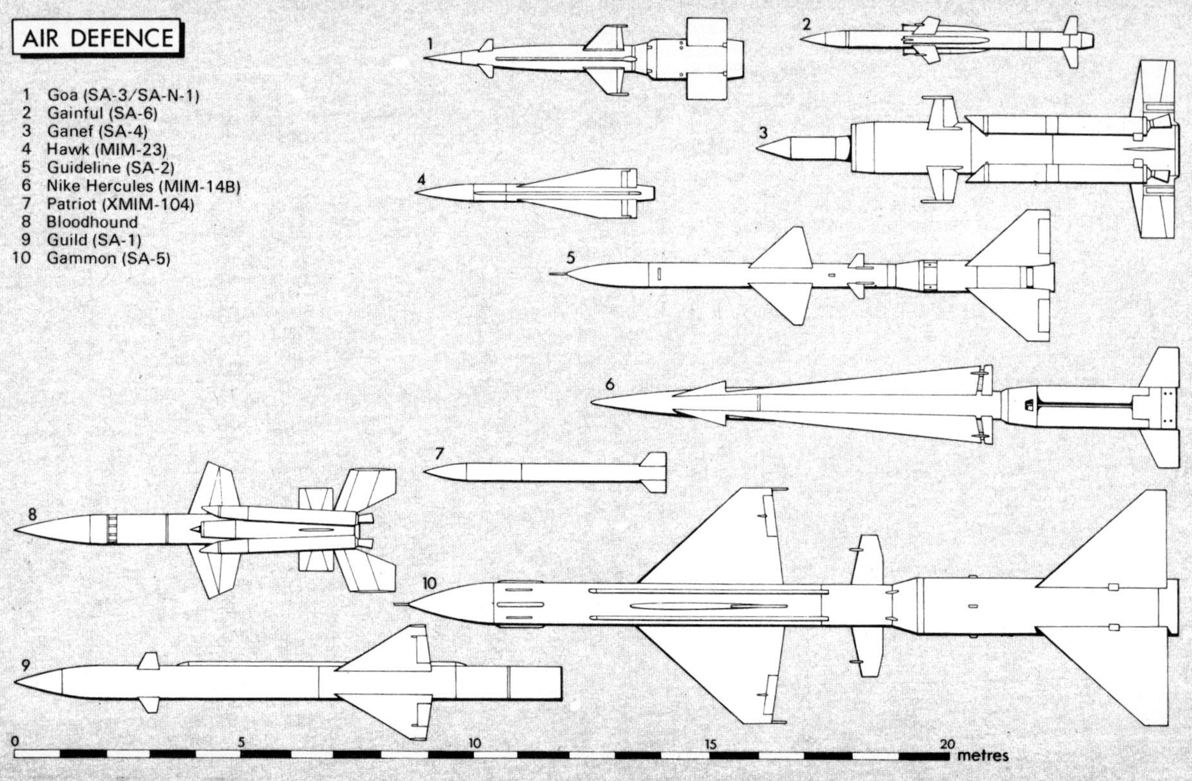
AIR DEFENCE
1 Goa (SA-3/SA-N-1)
2 Gainful (SA-6)
3 Ganef (SA-4)
4 Hawk (MIM-23)
5 Guideline (SA-2)
6 Nike Hercules (MIM-14B)
7 Patriot (XMIM-104)
8 Bloodhound
9 Guild (SA-1)
10 Gammon (SA-5)
1
2
3
4
5
6
7
8
9
10
0
5
10
15
20
metres

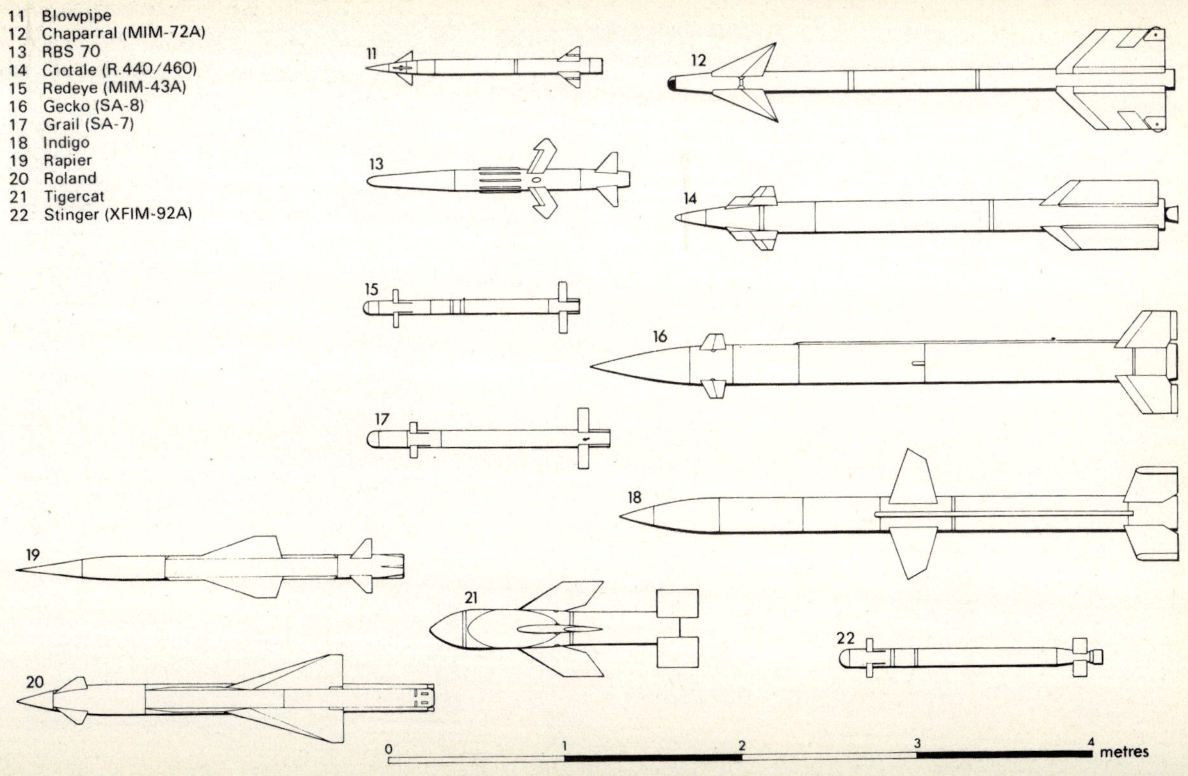
11 Blowpipe
12 Chaparral (MIM-72A)
13 RBS 70
14 Crotale (R.440/460)
15 Redeye (MIM-43A)
16 Gecko (SA-8)
17 Grail (SA-7)
18 Indigo
19 Rapier
20 Roland
21 Tigercat
22 Stinger (XFIM-92A)
11
12
13
14
15
16
17
18
19
21
22
20
0
1
2
3
4
metres

AIR DEFENCE MISSILES

The contents of this section are confined to land-based air defence missiles, from man-portable systems such as the Russian SA-7 (Grail) to ABM (Anti-Ballistic Missile) systems of which there are now Russian examples only since the United States decided to dismantle its ABM system (Safeguard) in October 1975. As a consequence of this decision the two missiles involved in Safeguard, Spartan and Sprint, no longer have a place in this pocket book. The USA is maintaining an ABM technology programme however, to ensure the ability to embark on a new system if future circumstances justify such a step.

Naval surface-to-air missiles will be found in the Naval Missile Systems pages. In some instances one basic missile will be used in both land-based and naval applications.

BLOODHOUND

Bloodhound Mk 2 is the successor to Bloodhound Mk 1 which first went into service with the RAF in 1958; development having been started in 1949. Bloodhound I was also exported in substantial numbers to Australia.

The first design study for the Mk 2 version was started in 1958; the new version went into service with the RAF in 1964. In the same year the equipment also went into service with the defence forces of Sweden and Switzerland. The Swedish and Swiss designations are RB68 and BL-B4, respectively. In late 1969 BAC were awarded a £10 million contract from the Singapore Government for the refurbishing and maintenance of the Bloodhound system previously operated by the RAF in that area.

The basic unit — a missile section — typically comprises four guided missiles and their launchers, a target illuminating radar (TIR) and a launch control post (LCP). The guidance principle is semi-active homing — the missile detecting and homing on the radiation reflected by the target when illuminated by the TIR.

In the mobile role for which it was first designed, the Bloodhound missile section is normally equipped with the Firelight TIR which is readily mobile and air-transportable. For static operations — or where system mobility is not a prime consideration — there is an alternative TIR, the Scorpion which is a larger radar giving a longer range. It can be broken down into transportable units but cannot be brought into or taken out of operation as quickly as the Firelight. Both radars are X-band CW Doppler equipments having good performance in the presence of natural or ECM interference. The majority of Bloodhounds now in service are deployed at either fixed or semi-permanent installations.

The RAF has Bloodhound sites in the UK and in West Germany as part of the NATO and UK Air Defence network. Bloodhound Mk 2 is also in service with the Swiss Armed Forces who operate about 64 launchers, the Singapore Air Force with 28 launchers, and the Royal Swedish Air Force.

Type: Surface-to-air
Configuration: Cylindrical body with pointed nose. Pivoted wings at mid-length in line with fixed horizontal tail surfaces. Ramjet sustainer motors above and below rear of fuselage. Four strap-on boosters each with large rectangular fin
Length: 8.46 m
Diameter: 54.6 cm
Span (Max): 2.83 m
Propulsion: Two Rolls-Royce Bristol Thor ramjets. Four jettisonable Bristol Aerojet solid boosters
Range: 80 km plus
Guidance: Semi-active radar homing. Twist-and-steer control
Warhead: High-explosive, proximity fuzed
Main Contractor: British Aircraft Corporation

Bloodhound air-defence missile leaves its launch ramp

BLOWPIPE (UK)

Blowpipe is a light, compact weapon for use in forward areas to give protection against low-level air attack. In its standard form Blowpipe is capable of full operational use by one man and is man-portable. The missile has subsequently been used in a variety of multiple launcher systems developed to provide both submarines and surface ships of all sizes with an air point defence capability. The former type of system is known as SLAM (Submarine/Surface Launched Air Missile), and is produced by Vickers Ltd Shipbuilding Group.

The missile is fitted with flares which in the early stages of flight are detected by a sensor in the Aiming Unit to gather the missile to the centre of the aimer's field of view. From then on the aimer guides the missile to the target by means of the controller with up/down and left/right movements. When within lethal distance of the target a proximity fuze in the nose of the missile detonates the warhead. Commands are transmitted to the missile over a radio link, to the missile control system. This works on twist and steer principles with one pair of nose-mounted control surfaces working differentially to produce roll, and the other pair producing lateral movements. A fully integrated IFF system is available for use with Blowpipe and is physically attached to the aiming unit. This facility prevents the inadvertent engagement of friendly aircraft which may have been incorrectly identified visually and considerably reduces reaction times.

Blowpipe was initially a private venture project by Short Bros & Harland Ltd, but it was adopted for full development after selection from three proposals put up to the Royal Radar Establishment in 1966 when a UK requirement was under study. Blowpipe is now in service with the British and Canadian Forces.

Type: Surface-to-air, or surface-to-surface, man-portable tactical missile. Submarine-launched version (SLAM)
Configuration: Very slender cylindrical body with long pointed nose. Delta cruciform control surfaces near nose. Cruciform tail fins
Length: 1.35 m
Diameter: 76 mm
Span (Max): 27 cm
Weight: 21.4 kg. (Launcher/sight with IFF and one missile, complete.)
Propulsion: Two-stage solid
Guidance: Radio command, optical tracking. Twist-and-steer control
Warhead: High-explosive, proximity fuze
Main Contractor: Short Brothers and Harland Ltd.

CHAPARRAL (MIM-72A)

The US Army Chaparral short-range battlefield air defence system uses the Sidewindeer 1C missile, modified for surface-to-air operation instead of its usual air-to-air role. Four missiles are mounted in ready-to-launch condition on a modified M730 tracked vehicle which also carries a supply of 12 missiles for reloading. The launch and control unit combines the launch assembly and the operator's optical sighting system. This unit can be used independently of the carrier vehicle, either in a fixed emplacement or as a trailer-mounted unit. The FAAR (Forward Area Alert Radar) can be used with the system to give increased detection range and IFF facilities.

Development of the Chaparral system began in 1965 with first trial firings in July of that year. Production was initiated for the US Army in 1968. In March 1974 a Pentagon plan to improve 1,500 Chaparral missiles in Fiscal Year 1975 was announced. Since then, new warhead, guidance and fuze components have been developed and the new standard model is the MIM-72C. Chaparral will be the US Army's main forward area low-level defence system until US Roland is ready in the 1980s.

Type: Surface-to-air. Close-range air defence system
Configuration: Slender cylindrical body with four large tail fins and four delta shaped control surfaces near rounded nose
Length: 2.92 m
Diameter: 12.7 cm
Span (Max): 63.5 cm
Weight: 84 kg
Propulsion: Solid Mk 36
Guidance: Optical aiming. infra-red homing
Warhead: High-explosive
Main Contractor: Aeronutronic Ford Corporation

*Blowpipe (*far left*) is man-portable weapon. Chaparrel system (*left*) uses a version of the Sidewinder missile*

CROTALE (R.440/R.460) (FRANCE)

Crotale is a short-range surface-to-air missile system for the all-weather interception of low-altitude targets. The complete weapon system can be mounted on wheeled vehicles, semi-mobile launchers or shipboard launchers and is air-transportable.

Mounted on wheeled vehicles, the complete system comprises up to three combined launch and command guidance vehicles and a surveillance radar vehicle. Four missiles, in containers, are carried by each launch vehicle, the missile being in ready-to-fire condition. Also carried by this vehicle is a monopulse fire-control radar capable of guiding two missiles simultaneously. Radar acquisition of the missile immediately after launch is aided by infra-red detection of exhaust heat: there is also an optical tracking device. Guidance signals are transmitted to the missile by radio.

On the second vehicle is a pulse-Doppler E/F-band surveillance radar having the good anti-clutter performance necessary for low-altitude target detection. Associated with the radar is an automatic target evaluation system that gives the weapon system its fast reaction time. The surveillance radar vehicle (called the Acquisition Unit) is able to provide command and control functions for one, two, or three launch vehicles (called Firing Units).

The system is also produced for South Africa, in which country it bears the designation of Cactus. Some differences between French and South African versions are probable but details are not known.

It was announced in 1975 that Thomson-CSF had been awarded a contract for the development and production of a mobile anti-aircraft system called Shahine for Saudi Arabia. Although based on the Crotale missile system, Shahine has some important differences. Instead of being mounted on a Hotchkiss-Brandt chassis, Shahine units are mounted on AMX 30 tank chassis, and the arrangement of equipment is different. Whereas a four-round launcher turret is standard for Crotale, the system for Saudi Arabia has one carrying six launcher-containers. Also, the radar (which is of a new type) is mounted on its own chassis. This enables a larger reflector to be used, thereby endowing the search radar with better discrimination. The launcher vehicles and their six Matra R.460 missiles each have their own tracking radar as in the original Crotale system.

The Naval Crotale was chosen for use by the French Navy in 1976 and is reported to have been selected by a number of other unspecified navies. This version employs an eight-round launcher turret, this also carrying radar and command link antennas, and a TV sensor.

Work started in 1964 and a prototype missile was completed in the following year. Overall responsibility for the development lies with Thomson-CSF group — while full responsibility for the missile lies with Matra. Series production of the system started up at the end of 1968 and deliveries to the South African forces began towards the end of 1971. The French Air Force ordered a quantity for part of an airfield defence network associated with strategic nuclear forces.

Crotale is in production for the French Air Force and is reported to be in service or on order for the forces of Libya, Kuwait, Pakistan, Spain and possibly other countries. Cactus is in service and is believed to be still in production for South African air defence. The Shahine version was due to begin tests by late 1977 and deliveries to Saudi Arabia were forecast for 1980. The French Navy has ordered the naval version of the system to equip several types of ship.

Type: Surface-to-air
Configuration: Slim cylindrical body with cruciform canard wings and four larger control and stabilisation surfaces at rear end
Length: 2.89 m
Diameter: 15 cm
Span (Max): 54 cm
Weight: 80 kg
Propulsion: Single stage solid
Range: 500 m - 8.5 km
Guidance: Radio command
Warhead: High-explosive, fragmentation with infra-red proximity fuze
Main Contractors: Thomson-CSF, Engins Matra

Shahine mobile air-defence system, shown here, uses a version of the Crotale missile, which also hasnaval applications and is known as Cactus in South African service

12

Gainful is both a command guided and a homing missile and is fully mobile, both missile launchers and fire control radar systems being mounted on separate PT-76 tracked vehicles. The fire control system is both sophisticated and, it would seem, effective. It comprises a primary search and acquisition radar, a target tracking and illuminating radar, a command link with secondary radar response for missile tracking and a missile-borne semi-active homing system. In ECM conditions some of the tracking functions can be performed optically.

The ground equipment of this complex is known by the NATO name Straight Flush. Its limitation would appear to be restricted search capability when operating without the support of other types of surveillance radar, such as the Long Track. The search/acquisition function is carried out in the 5-6GHz band, the target tracking/illuminating function in the 8-10GHz band and the command guidance and beacon response is also in the 8-10GHz band — though obviously not on the same frequencies. This sort of multiple frequency combination is of course not unlike that used in some Western missile systems of similar age.

The missile is constructed as a single-stage body containing a dual-thrust integral rocket-ramjet propulsion system. In the boost phase the missile is powered by a solid-propellant rocket motor which accelerates it at about 20g to about Mach 1.5. With the completion of this rocket burn the tail cone of the missile, which contains the booster nozzle, is jettisoned and the rocket propellant chamber becomes the combustion chamber for the ramjet, ram air being supplied through four intakes disposed symmetrically around the centre section of the missile. The ramjet takes the speed up to almost Mach 2.8.

Two sets of cruciform fins at the centre and tail of the missile provide stability and aerodynamic control, pitch and yaw being controlled at the centre and roll at the tail. The tail fins also carry the command link receiver aerial and the beacon transmitter aerial. The nose of the missile is an ogival radome for the homing head.

In addition to the Soviet forces, the SA-6 Gainful is known to have been supplied to the armed services of Egypt, Libya, Syria, Vietnam, and possibly to others. According to the Indian press, it was reported in 1976 the arrangements had been made for its manufacture in India. The first public showing of Gainful was in Moscow in 1967.

Type: Surface-to-air, tactical anti-aircraft weapon
Configuration: Slender cylindrical body with pointed nose. Small cruciform wings near mid-length and cruciform tail fins. Four air inlet ducts between wings
Length: 6.2 m
Diameter: 33 cm
Span (Max): 1.24 m
Weight: 550 kg
Propulsion: Solid booster, air-breathing solid sustainer
Range: Up to 60 km
Guidance: Command plus semi-active radar homing
Warhead: High-explosive. Infra-red fuze

SA-6 Gainful mobile air-defence missiles

GALOSH (ABM-1)

Galosh is the NATO code name for the Soviet anti-ballistic missile, first 'shown' to the public in 1964 and initially known as SA-7 in the US alphanumeric code. The emergence of the man-portable Grail (SA-7), and the consequent confusion arising from there being two SA-7s, led to the Galosh becoming the ABM-1. Since the missile has so far only been thus 'shown' housed in an almost all-enveloping container little is known about it, other than that it has four first stage nozzles and that the container is some 20m long and has an internal diameter of about 2.75m. Soviet pictures purporting to show the launch of a long-range ABM Missile that could well be Galosh have been seen, but they convey little information other than that the missile is more or less conical in shape. In official American defence circles Galosh has been said to more closely resemble Spartan than Sprint, of the former US counterpart ABM system, Safeguard. Long-range operation, implying exo-atmospheric intercept was also reported.

It is believed that Galosh carries a multi-megaton warhead suitable for exo-atmospheric missile interception and that it has a range of at least 300 km. On this assumption it is somewhat inferior to the US Spartan missile.

Operational at Moscow in four sites containing a total of 64 launchers. The Moscow ABM system (ABM-1) consists of battle management radars, engagement radars and Galosh launchers. Eight complexes were originally under construction, but only four, with 64 launchers, have been completed. Each complex consists of two large tracking radars, four smaller interceptor guidance and tracking radars, and 16 missile launchers. Radar types include Triad, Dog House, Chekhov, and Hen House. Official DoD statements have referred to possible improvements to the radars associated with the Moscow ABM complex, and the Russians are reported to be building two 'massive' OTH radars facing the United States.

The Russians are reputed to have developed an improved version of the Galosh ABM missile. Unofficial sources credit this missile with the designation SH-4, but this has not been confirmed, nor has its origin been disclosed. An important feature of the new development is its alleged 'loiter' capability. At first, it was thought that this added capability was simply a way of improving the missile's exo-atmospheric interception capability. Recent reports of a fairly detailed nature, however, suggest that the new missile differs in function from the earlier one as well as in performance. Whereas the basic Galosh missile is believed to be a high-altitude exo-atmospheric interceptor, the new version — while still retaining a slant range capability of some 350-700km — can stop and start its motor four or five times at upper altitudes — while decoys are sorted from warheads by the ground radars — and make a high-altitude interception in the re-entry vehicle's terminal phase. Flight tests have been reported from American sources but otherwise the status is unknown.

Galosh ABM missiles in their transit containers in Red Square

GAMMON (SA-5) (USSR)

Griffon was widely accepted as the NATO code name for a large surface-to-air weapon first displayed in public in 1963, but the practice in official US defence circles has recently been to designate this missile as the SA-5, Gammon. It has been variously described as an unmanned long-range interceptor, an anti-aircraft missile and an anti-missile missile, but it seems probable that although it may have some anti-missile capability it is primarily suitable for long-range anti-aircraft operations.

The nose of the missile houses a radar reflector of at least 60 cm diameter which can be combined with an active radar target seeking system. Generally similar in size and weight to the US Nike Zeus missile, Gammon is thought to be somewhat inferior in performance and to have a lesser anti-missile capability. The fact that it is evidently manoeuvred aerodynamically indicates that its ability to home on to targets travelling at missile speeds must be very limited. Although the missile evidently has two stages, it has been suggested that it may in fact have three — the third being a motor built into the warhead section and used to power the warhead and homing system during the final stages of interception. This, coupled with some form of thrust vector control could materially enhance any anti-missile capability.

Operational since 1967, Gammon provides point defence for a number of major areas of strategic importance. The current Soviet deployment plan appears to be based on use of Gammon as the high-altitude missile and the SA-3 Goa as the low/medium altitude weapon for the strategic air defence of the Soviet Union. Numbers of both types are believed to be increasing steadily and in the region of 1,000 Gammons are thought to be deployed, some of them in association with ABM-1 Galosh missiles in the Moscow missile defence system.

Type: Surface-to-air, strategic air defence missile
Configuration: Cylindrical body with booster motor of larger diameter. Very large cruciform fins on booster and main section of missile, smaller cruciform indexed at 45 degrees to other surfaces
Length: 16.5 m
Diameter: 100 cm booster: 80 cm second stage
Span (Max): 3.65 m
Weight: 10 000 kg Approx
Propulsion: Two- or three-stage solid. It is not known for sure whether or not the payload section of the missile contains a third propulsiion motor
Range: 250 km Estimated
Guidance: Probably command plus active radar homing
Warhead: Possibly nuclear and high-explosive versions exist

GANEF (SA-4)

Ganef is the NATO code name for a surface-to-air missile launched from a tracked vehicle that was first seen in public in Moscow in 1964. Its US alpha-numeric code is SA-4. Two missiles are mounted on the armoured launcher from which they can be raised into the firing position. The system is thus highly mobile and can be brought into action rapidly: it can also be airlifted by such aircraft as the An-22 heavy freighter. The use of an armoured launcher suggests operation in forward areas and it is believed that the missile can also be used in a ground-to-ground role. For target acquisition and fire control a separately mounted G/H-band radar whose NATO code-name is Pat Hand is used. Longer range surveillance is usually provided by Long Track radars which are also used with other mobile Russian missiles such as the SA-6 Gainful. Some observers claim to have detected a later model of Ganef, but firm evidence is sparse.

Ganef is widely deployed in Russia. It was also at one time deployed in Egypt, but since it appears not to have been used in the 1973 Arab-Israeli war it seems likely that the missiles were withdrawn — possibly in exchange for the SA-6 Gainful missiles which were used in the war. It is not known to have been supplied elsewhere outside the USSR.

Type: Surface-to-air
Configuration: Cylindrical body with main section of larger diameter than payload and guidance package carried in annular ramjet air intake. Cruciform wings and tail surfaces, set at 45 degrees. Four boosters aft of wings
Length: 9 m
Diameter: 90 cm
Span (Max): 2.6 m
Weight: About 1 800 kg
Propulsion: Four solid boosters. Ramjet sustainer
Range: About 70 km
Guidance: Command
Warhead: High-explosive

SA-4 Ganef medium-range mobile air-defence missiles

GASKIN (SA-9) (USSR)

The SA-9 Gaskin designation refers to what is apparently the most elaborate of the several vehicle-mounted derivatives of the man-portable SA-7 Grail short-range anti-aircraft missile. In this case the vehicle is the amphibious BRDM-2 which carries a mounting comprising four launcher containers with an aimer's position incorporated in the base. When operated in an autonomous mode, the method of target acquisition is presumed to be by visual means with subsequent aiming of the weapons by optical systems. The remainder of the engagement procedure is probably similar to that for the man-portable SA-7, with indicator lights to denote missile seeker operation and target acquisition. When the latter has been achieved the operator is free to fire. It is not known what, if any, IFF provisions are made.

When used as part of a larger air defence column, SA-9 vehicles can be linked to search radars to assist in target acquisition, and another method could be that of using radio communications links to 'tell off' targets to individual SA-9 firing units. Gaskin firing units are also deployed in forward battle areas with ZSU-23-4 radar-directed AA gun vehicles.

Little is known for certain about the SA-9 missile itself, but it is claimed to be basically an improved SA-7, the nature of the specific improvements varying with individual observers' views. It is generally assumed that these include a larger, more powerful warhead, improved propulsion, and perhaps different aerodynamic control surfaces for greater manoeuvrability. A UK MoD report issued in April 1976 quoted the approximate range as 8 km. So far as it has been possible to establish, the SA-9 Gaskin is so far deployed only with Warsaw Pact forces.

SA-9 Gaskin air-defence system uses a version of the man-portable SA-7 Grail missile

GECKO (SA-8)

This recently revealed system (November 1975, Red Square, Moscow) has been dubbed the Soviet Union's Roland by virtue of the apparent similarity of the underlying operational philosophies of these two autonomous, highly mobile forward air defence systems. Equally obvious was the strong possibiliy that Gecko and the naval close-in missile known as SA-N-4 shared common origins. Not that identifiable pictures of the latter's 'pop-up' missiles had been made public to compare with those of SA-8 photographed in Moscow. Instead it is the associated compact but complex radar groups of the two missile systems which afford the link.

The SA-8 is a command guidance system, and in addition to antennae for the search and tracking functions, the radar group features two separate beacon tracking antennae and two transmitting horns for the command guidance signals. The main target tracking radar is a Cassegrain type array, possibly using conical scan and a pulsed transmission waveform, and the two smaller Cassegrain antennae flank the main tracker. They are mounted to provide a limited degree of movement relative to the main array and are assumed to be missile trackers. The command link horns for each are mounted beneath the missile tracker antennas.

This arrangement indicates provision for simultaneous firing of two missiles with separate guidance for each. One possible use of this facility would be to enable simultaneous engagement of two separate targets within a formation, and another is to allow for 'split engagements' with one target being tracked by the radar and another by the telescopic optical tracker which is incorporated in the system. There are other possibilities but it seems certain that a multiple target capability exists.

Both the twin-missile launcher arms and the search radar antenna fold down for reduced transit height. The nature of the reloading arrangement cannot be ascertained from the evidence so far available, but the hull of the six-wheeled vehicles built to carry the SA-8 should be able to accommodate at least one full reload of four missiles. Two versions of the carrier vehicle have been noted, both about 9 m long but differing slightly in overall length.

The missile is estimated to be 3.2 m in length and 21 cm in diameter. Control is by four small canard surfaces and there is a cruciform tail assembly of about 60 cm span. The design is probably optimised for high acceleration, maximum speed, and manoeuvrability rather than range. Maximum range is not likely to be greater than 8-16 km, while the speed could be in the region of Mach 2.

Issues of Gecko so far are thought to be limited to Soviet formations.

GOA (SA-3/SA-N-1) (USSR)

Goa is the NATO code name for an anti-aircraft guided missile for use both by the Soviet Army and by the Soviet Navy. Small enough to be carried in pairs on the ZIL-157 vehicle that is used as a tractor for the trailer transporters of both Guideline and Guild, Goa is a two-stage missile that is thought to have been intended for much the same operational role as the US Hawk — that is to say, short-range defence against low-flying targets.

Control of the missile in flight is by movable foreplane surfaces, the rear mounted wings of the second stage being fixed. Considering the operational role suggested above it seems possible that the missile has some form of homing device in the nose; if so, the positioning of the control actuators in the same part of the missile would make a compact guidance package. Basically, however, it seems fairly definite that Goa is a command-guided weapon.

Goa is used in conjunction with an X-band fire control radar whose NATO code-name is Low Blow and is commonly associated with an acquisition radar code-named Flat Face. In the linked installation of Goa and Guideline (SA-2) missiles built by the Egyptians, with Russian assistance, near the Suez Canal in 1970, the Goa missiles were mounted on substantial trainable launchers, contrasting with the simple mobile ramps from which they are normally launched. Tracked vehicle mountings have also been seen. In the first few months of 1976 a new quadruple launcher for the SA-3 was observed in Yugoslavia. It has not been established if this development is of Yugoslavian origin or if similar launchers are deployed elsehwhere.

In addition to its land-based uses Goa is extensively deployed on ships of the Soviet Navy and Polish Navy, in which cases the US designation is SA-N-1.

Introduced into service in 1961 and now widely deployed. The number in use is expected to show a further increase in the next few years. In January 1976 the US Secretary of Defence reported that additional SA-3 sites had been deployed in the Soviet Union. Goa has been supplied to a number of other states, among them Egypt, Iraq, Libya, Syria, Vietnam, Uganda, and Yugoslavia. It can be expected to increase both within and beyond the Warsaw Pact.

Type: Surface-to-air. Land-based and ship-borne versions
Configuration: Circular cross-section main body having gradual taper to pointed nose. Cropped delta cruciform wings and foreplanes. Larger diameter booster motor has large rectangular cruciform fins
Length: 6.7 m with booster
Diameter: 45 cm. Booster: 61 cm
Span (Max): 1.22 m
Propulsion: Two-stage solid
Range: 25-30 km
Guidance: Probably radar beam-rider, with semi-active radar homing
Warhead: High-explosive

The SA-7, Grail (once widely known as Strela) is an extensively deployed man-portable infra-red homing light anti-aircraft missile, very similar in concept to the American Redeye but preceeding the latter into service. It relies upon a tail pursuit interception to engage low-flying aircraft targets and has proved especially effective against helicopters. Overall length of the weapon is 1.5 m and the weight is in the region of 15 kg. An official UK MoD report issued in April 1976 stated that the range of the SA-7 is approximately 9-10 km. Simple optical sighting and tracking is employed, the I/R seeker being activated when the operator has acquired the target. An indicator light denotes seeker acquisition and the operator is then free to fire the missile. Flares were for a time found to provide an effective counter-measure but it is reported that later models are equipped with filters to combat this tactic. The existence of an improved, Mk2 version of SA-7 has been claimed, the principal difference lying in a boosted propellant charge to increase range and speed of the missile.

Over the years that SA-7 has been in use — probably at least eight — additional forms of deployment have been evolved, apparently by both Russia and individual client states. In October 1975, for example, a number of Jeep-type vehicles were seen during an Egyptian military parade with a rudimentary platform and support for a SA-7 gunner at the rear of the vehicle. Four reload rounds were attached to the sides of the vehicle.

The Soviet development is the SA-9, Gaskin, according to the British report mentioned above. In this version four SA-7 missiles are housed in four rectangular launcher boxes, turret-mounted on a BRDM AFV. Some observers claim that the missiles used in this system are larger and heavier and with correspondingly higher performance than the SA-7 Grail, but no evidence either way has been obtained. The MoD report, in fact, quotes a slightly lower range value for the SA-9 Gaskin missile than for the basic SA-7.

In widespread use with Warsaw Pact nations, Egypt, Syria, Vietnam, Yemen, and probably others including, Cuba, China, Morocco, and Yugoslavia.

GUIDELINE (SA-2)

Guideline is a medium-range, anti-aircraft guided weapon system, known also by the US as alpha-numeric code SA-2, which is standard equipment in the Soviet forces and which has been exported in large numbers to many countries outside the Warsaw Pact territory. Guideline missiles were used against US B-52 bombers during the bombing raids on North Vietnam prior to the peace settlement. Initially they scored a number of successes but subsequently it appeared that the B-52 ECM equipment was more than a match for such ECCM equipment as the North Vietnamese had available.

The Soviet designation for the missile — or one version thereof — is believed to be V750VK and that of the complete weapon system — including radar and power supplies — V75SM. The system is land-mobile the missile being mounted on a Zil-157 cross-country semi-trailer transporter-erector. Various models of the missile have been observed over the years exhibiting relatively minor changes in external appearances. The details given below relate to a type that was supplied to Egypt, but it is known that later versions exist. In particular, one that was seen in Moscow in 1967 is somewhat longer and has a larger warhead.

The radio command guidance system is believed to be quite straightforward. Targets are tracked by radar which feeds data to a computer: this in turn produces signals which modulate the output of the command transmitter. Two sets of four strip antennae are mounted fore and aft of the missile wings to receive these signals.

Entry into Russian service was in 1958, but numbers are now declining in the USSR. Soviet client countries reported to have received the SA-2 Guideline include: Afghanistan, Albania, Algeria, Bulgaria, China, Cuba, Czechoslovakia, Egypt, East Germany, Hungary, India, Iraq, North Korea, Libya, Poland, Romania, Vietnam and Yugoslavia. Some of these nations may well have introduced modifications of varying degrees to suit their own purposes, and China is thought to have established its own production line at one time.

Type: Surface-to-air, land-based and ship-borne versions
Configuration: Slender cylindrical body with pointed nose. Tail end of second stage slightly flared to match diameter of booster motor. Cruciform wings and tail surfaces. Different marks exist with and without small nose fins of various shapes
Length: 10.7 m
Diameter: 50 cm. Booster: 70 cm
Span (Max): 1.7 m
Weight: 2 300 kg
Propulsion: Solid booster, liquid sustainer
Range: 40-50 km
Guidance: Radio command. Semi-active radar homing probable on some versions
Warhead: High-explosive, 130 kg. Contact, proximity, and command fuzes reported

SA-2 Guideline deployed for action at a prepared site

GUILD (SA-1) (USSR)

Guild is the NATO code name for an anti-aircraft guided missile system first shown in Moscow in 1960 and identified by the number SA-1 in the US alpha-numeric code.

About 12m long and 70 cm in diameter, this missile has no separate booster stage and is said by official sources to be liquid-propelled. Movable cruciform foreplane surfaces are another feature of the missile that distinguishes it from Guideline, which has movable tail surfaces but which is otherwise generally of much the same size as Guild. A British MoD report of April 1976 mentions a range of 32 km for the SA-1. Although seen on transporters, US Defence officials classify this weapon as part of the Soviet strategic air defence force. First deployed in 1954 and still in service, but expected to decrease in numbers because of obsolescence.

Strategic air-defence Guild (SA-1) has been said to be obsolescent for some years but remains in service (left). American Hawk (right) is widely used in various forms. Trailer carries the tracker/illuminator radar

HAWK (homing-all-the-way-killer), MIM-23A, is a surface-to-air missile designed primarily to engage low-level supersonic targets but capable of intercepting such targets at altitudes ranging from 30 to 11,000 m.

The missile also has an anti-missile capability — indeed the first known kill of one supersonic missile by another was achieved in January 1960 when a Hawk destroyed an Honest John battlefield missile. A self-propelled tracked launcher vehicle is generally used, although a trailer-type launcher exists as an alternative. Both types are three-round launchers. The same number of missiles is carried by the tracked loader vehicle, which is one of several that comprise a Hawk battery. Both pulse and CW target acquisition radars are used, with a separate range-only radar to provide range data for the latter.

Hawk development began in 1954 and the system entered service

in 1959. In the latter year a European consortium was formed to produce Hawk missiles in Belgium, France, Italy, Netherlands, and West Germany. Licence manufacture is also undertaken in Japan and is expected to continue until 1977, although the European production programme is stated to have been completed.

Full production of a new model, Improved Hawk, MIM-23B, was authorised in January 1972. This recent development has a new guidance package, a larger warhead and an improved solid propellant. Other improvements include increased ECM protection and the introduction of the 'certified round' procedure whereby missiles are delivered ready to fire without field maintenance or testing. Improved Hawk is currently in production: the first Basic Hawk battalion was converted to Improved Hawk in November 1972 in USAREUR/Seventh Army. Hawk and Improved Hawk missile systems are widely deployed throughout the world, and in the US Army and USMC it will remain the mainstay of field air defences until replaced by the Patriot, MIM-104.

The list of countries operating Hawks is a long one: Belgium, Denmark, France, West Germany, Greece, Iran, Israel, Italy, Japan, Kuwait, the Netherlands, Saudi Arabia, South Korea, Spain, Sweden, Taiwan, Thailand, and the USA. All but seven (Belgium, Israel, Japan, South Korea, Sweden, Taiwan, and Thailand) operate or have ordered the Improved Hawk system.

Type: Surface-to-air
Configuration: Slender cylindrical body, with slight taper at tail end and pointed nose. Cruciform long-chord cropped delta wings with full-span control surfaces at trailing edges
Length: 5.03 m
Diameter: 36 cm
Span (Max): 1.2 m
Weight: 587 kg
Propulsion: Two-stage solid
Range: 35 km
Guidance: Semi-active CW radar
Warhead: High-explosive, blast fragmentation type
Main Contractor: Raytheon Company

INDIGO

Indigo is a short-range surface-to-air land-based weapon system. It has undergone an extensive series of successful trials at the Italian firing range in Sardinia and is now fully operational with the Italian Army in a towed version. This comprises one or more Indigo trailer launchers and associated trailer power units together with a modified Superfledermaus fire control system and LPD/20 pulse doppler search radar, both of which are made by Contraves. The modification to the Superfledermaus consists mainly of the addition of missile engagement and guidance computers, a missile control panel, a command transmitter and an infra-red tracker.

A self-propelled version is being developed for the Italian Army. This will have an Officine Galileo fire control system and Thomson-CSF Eldorado/Mirador radars. Two tracked vehicles will be used, one carrying the missile launcher and the other the radars and fire control system. The complete system is known as the MEI air defence system.

Operation of the missile is similar in principle to that of the Sistel Sea Killer Mk 1 and 2 missiles described elsewhere in this book. Primarily intended as a beam-riding system, the missile can also be controlled over the radio command guidance link with optical and infra-red tracking if the beam-riding function is rendered inoperative by jamming or otherwise. Unlike the Sea Killers, however, there is no radio altimeter in the system

Development of the system was initiated by Contraves Italiana SpA in 1962 and the first prototype was completed in 1963. In 1969 the project was taken over by Sistel SpA.

Type: Surface-to-air. Mobile anti-aircraft system
Configuration: Cylindrical body with pointed nose. Cruciform wings at mid-length and rectangular cruciform tail fins
Length: 3.07 m
Diameter: 19.5 cm
Span (Max): 81 cm
Weight: 120 kg
Propulsion: Solid, 3,800 kg static thrust
Range: 10 km
Guidance: Beam riding/radio command, with standby optical and infra-red tracking with radio command
Warhead: High-explosive fragmentation, 22 kg, with infra-red proximity fuze
Main Contractor: Sistel SpA

NIKE HERCULES (MIM-14B) (USA)

Having served as the mainstay of American air defence missile forces for almost 20 years, Nike Hercules is progressively being phased out of service but this process is not likely to be completed in the USA for some years. Its use by eight other NATO and SEATO countries is likely to extend still further into the future. The non-nuclear warhead version is still in production under licence in Japan.

Development of Nike Hercules was started in 1953, the year its predecessor Nike Ajax entered operational service, and in 1960 it attracted attention by successfully intercepting another missile at a range of 48 km and at a height of some 30,500 m.

Major improvements were made to the system in 1961 when new radars and other modifications were incorporated. The number of active batteries has steadily declined since 1963 when there were 134 batteries. In February 1974, the US Army disclosed that 48 Nike Hercules batteries within the continental USA would be phased out by the end of that year, leaving operational batteries in Alaska and Florida, and training units at Fort Sill.

Type: Surface-to-air, strategic defence missile
Configuration: Slender cylindrical body with pointed nose, small cruciform canard fins, long-chord cruciform delta wings with prominent trailing edge control surfaces. Larger booster motor assembly comprising a cluster of four rockets and cruciform fins
Length: 12.5 m
Diameter: 80 cm
Span (Max): 2.13 m
Weight: 4 500 kg
Propulsion: Two-stage solid
Range: 140 km plus
Guidance: Command
Warhead: Nuclear and high-explosive types available
Main Contractor: Western Electric Company

*Nike Hercules (*left*) has served as main air-defence missile for the USA and some allied countries for almost 20 years. Newcomer on the scene is Patriot, formerly SAM-D, (*right*)*

PATRIOT (XMIM-104)

Patriot (XMIM-104) is an advanced surface-to-air missile system intended as a replacement for the American Hawk and Nike-Hercules systems in the 1980s, and now in the development stage. The project was formerly known as SAM-D (Surface-to-air Missile Development), which was the successor to two earlier study programmes managed by the US Army Missile Command's Research and Development Directorate. These studies were concerned with systems known as the Field Army Ballistic Missile Defence System (FABMDS) and the Army Air Defence System for the 1970s (AADS-70s). In January 1965 the SAM-D requirement was specified and a study was directed towards defence against high-performance aircraft — at all altitudes — and short-range missiles. SAM-D was placed under project management at the US Army Missile Command in August 1965. In February 1970 the missile successfully completed its first launch environmental test.

Patriot can be developed as a battery to provide circular defensive coverage or as a fire section to provide coverage over a sector. A fire section will consist of one fire control group and one or more (normally two) launcher groups, and may be detached from the major control elements for autonomous operations. A battery will normally include four firing sections, and each launcher group will contain six missiles, thus giving each battery a total of 48 missiles. Such a field battery would be mounted on approximately 12 vehicles and would include three main elements: fire control, launchers, battery control, and communications groups.

The fire control group will contain the phased array radar, a radar/weapons control computer, communications and prime power equipment mounted on a single tracked or wheeled vehicle. It will also house the operators, controls and displays for the firing section. The radar/weapons control computer will control the phased array radar in its search, acquisition, track and engage functions. It will also assess the threat and determine the best engagement procedure. The phased array radar antenna will have an eight-foot diameter face made up of 5240 phase shifters. This radar will be capable of performing all the functions for which several radars are needed in some other systems — notably, of course, Hawk which can use four radars at a time and Nike Hercules which can use five.

The battery control group will co-ordinate firings within a battery and serve as a communications centre. It will house a computer for handling high data-rates and processes and will co-ordinate information between radars and pass on fire-control information. Missile guidance is by a novel combination of command link and TVM (Tracking Via Missile) in which data obtained by the missile's own radar head is relayed to the ground to enhance terminal guidance.

Tests carried out in 1975 yielded 12 successes from 14 firings and following the formal decision to proceed with full scale development in August 1976 an ambitious series of trials involving target drones protected by manned electronic jamming aircraft was planned for 1977-8.

Type: Surface-to-air, land mobile
Configuration: Cylindrical body with pointed nose. Cruciform cropped delta tail surfaces
Length: 5.18 m
Diameter: 41 cm
Weight: Classified
Propulsion: Single-stage solid TX-486
Range: Classified
Guidance: Command with TVM (Tracking Via Missile) homing
Warhead: Nuclear or high explosive
Main Contractors: Raytheon Company, system. Martin Marietta, missile

RAPIER

Rapier is a mobile, quick-reaction, low-level air defence missile system for use in forward areas or for point defence. It was developed to satisfy a British defence requirement and was the subject of a design study in 1963. The resulting system which first entered service with the British Army comprised a trailer Fire Unit towed by a long-wheelbase Land Rover which also carries the optical tracker and four missiles in sealed containers, and a second Land Rover which tows a missile re-supply trailer. Later developments of the Rapier system include alternative mounting arrangements (on AFVs etc) and a Rapier Blindfire radar (DN-181) which can be added to provide an all-weather and 'dark-fire' capability. Blindfire radars are being supplied to the armed forces of Abu Dhabi, Australia, Iran and the UK.

In addition to the missile launch mechanism, the Firing Unit contains a search and acquisition radar which works in conjunction with an IFF interrogator-responser, a microwave command transmitter and a computer which serves all parts of the system. When the surveillance radar detects a target the IFF automatically interrogates it to determine whether it is friendly or hostile. If the target is friendly all data on it is cancelled and the radar continues its search; if it is hostile the Tracker operator is alerted, the radar data is used to direct the Tracker towards the target and the operator will see it in his optical sight.

After acquiring the target optically the operator takes over from the radar and tracks it using a joy-stick control, the launcher having been automatically aligned so that the missile will fly on the line of sight to the target. A television camera in the Tracker collimated to the tracking telescope detects flares mounted in the missile tail and measures any deviation from the sightline. These measurements are then fed to the fire unit computer which then causes the command transmitter to transmit correction signals to the missile.

Rapier is fully operational with the British Army and RAF Regiment, with units based in both Western Europe and the UK; the Imperial Iranian Air Force; Zambian Army; and is being supplied to the Imperial Iranian Army in tracked form; the Abu Dhabi Defence Forces; Omani Armed Forces; and Australian Army. By early 1977, total exports of Rapier systems exceeded £600 million and joint production facilities had been set up in Iran for the manufacture of Tracked Rapier.

Type: Surface-to-air. Close-range anti-aircraft missile
Configuration: Slender cylindrical body with pointed nose and cruciform cropped delta wings near mid-length. Small cruciform control surfaces near tail
Length: 2.24 m
Diameter: 13 cm
Span (Max): 38 cm
Weight: Classified. Believed about 43 kg
Propulsion: Two-stage solid
Range: Classified
Guidance: Command to line-of-sight. Optical tracking with optional 'add-on' Blindfire radar tracker
Warhead: High-explosive with crush fuze
Main Contractor: British Aircraft Corporation

RBS 70 (SWEDEN)

RBS 70 is a portable surface-to-air missile system operating on the beam-riding principle, and a laser beam is used rather than one of non-coherent light, which gives it substantial immunity to jamming. It is suitable for use against both fighter-bombers and helicopters, has a range of about 5 km and can be operated by one man — but for continuous firing a second man is needed. The system is suitable for integration with other surface-to-air weapons or as a separate unit. RBS 70 can utilise precision target information from a search radar, and it is both possible and desirable to group several RBS 70 units with one or more search radars to form an anti-aircraft battery. RBS 70 is also designed to utilise IFF signals. A mobile G/H-band radar has been developed by LME for use with the RBS 70, the PS-70/R, now known under the name of Giraffe.

Three packs — stand, sight and missile in container — make up the complete firing unit which can be assembled and ready to fire in less than 30 seconds. The stand has three legs and forms the suspension device for the telescopic sight which has gyro-stabilised optics and which also contains the laser transmitter.

To slew on to the target the operator uses coarse aiming, turning the whole sight housing in the required direction. For tracking, a thumb lever is used to operate the gyro-stabilised optics.

The missile is launched from a container which also serves for both storage and transport. A starting motor ejects the missile from this launch container and separates from it at the mouth of the tube after which the missile sustainer takes over. The missile carries a receiver for the guidance beam and a small computer for calculation of control signals for the control surfaces. The warhead has both proximity and percussion fuzes.

Development was started in 1969 by AB Bofors under the auspices of the Swedish Defence Materiel Administration, and completed in 1974. Production for the Swedish Forces was approved in 1975 and deliveries have begun. The Swiss Government assisted in the funding of RBS 70 development and is expected to approve procurement of the system. Demonstrations have been given to at least 16 countries. A naval version has been considered.

Type: Surface-to-air. Man portable
Configuration: Tube launched
Length: 1.32 m
Diameter: 10.6 cm
Weight: 20 kg approx
Propulsion: Solid. Ejector motor remains in launch tube
Range: 5 km
Guidance: Laser beam-rider. Target tracking by stabilised optical system
Warhead: High-explosive. Active optical proximity fuze
Main Contractor: AB Bofors, Sweden

Tracked Rapier (far left) *is being produced for Iran and exports exceeded £600 million in early 1977. Sweden and Switzerland contributed to development of RBS-70* (left)

REDEYE (MIM-43A) (USA)

Redeye is employed in forward battle areas to protect combat troops from low-level aircraft, and for this role is man-portable and shoulder-launched. On sighting an attacker, the operator tracks the target with the optical sight incorporated in the launcher, at the same time energising the missile guidance system. A vibrator/buzzer in the launcher gripstock indicates to the operator when the missile is ready to be fired. After launch the missile is guided to its target by the infra-red homing system.

Development began in 1959 and initial production was announced in 1964, since when Redeye production reached peak rates of more than 1,000 per month. Production against US requirements ceased in 1970. In addition to the US Army and Marine Corps, users of Redeye include Australian and Swedish forces. A successor is the Stinger missile (which see).

Type: Surface-to-air. Man-portable anti-aircraft weapon
Configuration: Cylindrical body with rounded nose. Flip-out cruciform tail fins and nose vanes. Tube-launched
Length: 1.2 m
Diameter: 76 mm
Weight: 13 kg
Propulsion: Two-stage solid
Range: Classified. Probably about 3 km
Guidance: Optical aiming. Infra-red homing
Warhead: High-explosive
Main Contractor: General Dynamics Corporation

Redeye is optically aimed and uses infra-red homing

Roland is a surface-to-air missile system for the defence of armoured columns and similar mobile fighting units against low-level air attack. Land-based and ship-board versions have been designed, and the former category is produced in two forms for clear-weather and all-weather operation, and known as Roland I and Roland II, respectively. For information on the naval Roland, see the following section of this volume.

Arising from the AFV inventories of the two Armies responsible for the original Roland programme, France and West Germany, fitting and deployment of the Roland system involves two tank chassis types, the AMX-30R and the SPZ Marder. For France, the AMX-30R will be fitted with both Roland I and II systems, whereas the German requirement is for the all-weather Roland II only.

A third version of the system (US Roland) is the result of the American decision in favour of Roland following an international competition for a weapon to meet the US Army SHORADS/LOFAADS (Short Range Air Defence System/Low-level Forward Area Air Defence System) requirement. The choice of carrier vehicle for the US Roland has not been finalised and among other changes is the use of a different tracking radar. A further possibility is a special version for the Belgian Army which would have the firing unit carried on a Berliet GBD 4 x 4 wheeled vehicle instead of a tracked one, and with a Crotale search radar in place of the standard Roland equipment. This variant is provisionally designated Roland IIS.

The complete system is carried on a tracked launch vehicle and the Roland I configuration comprises a target acquisition and

Roland II on an AMX 30 tank

identification radar sub-system, an optical infra-red tracking and guidance sub-system and a launcher with missiles. The launcher is loaded from magazines of missiles carried in the vehicle by a hydraulic system that lifts a new missile into position each time one is fired. In the Roland II configuration all the sub-systems listed above are retained so that the system can be operated in exactly the same manner as the first — but an accurate target tracking radar is added to enable the system to be used in visibility conditions that make optical target tracking impossible. Additionally, the possibility of switching from automatic to optical tracking even while the missile is in flight provides a useful counter-countermeasures facility.

The present Franco-German production plan calls for a total of 252 firing units for France and 394 for Germany. The French total includes 144 Roland I firing units on AMX-30Rs; 26 spare Roland I firing units; and 12 spare Roland IIs. Germany will have 340 Roland II firing units plus 54 spare units. The production of Roland munitions (missiles in their tubes) allows for France to receive 7,300 Roland I and 3,500 Roland II, and Germany to have 12,200 Roland II rounds. Introduction to operational service with the French and German Armies is expected to begin during 1977 and continue until well into the 1980s.

The Brazilian Army decision to adopt the Roland system was announced in October 1972, but little more has been heard since then and the future of this programme is in some doubt. Late in 1975 the Norwegian Government announced its decision to purchase Roland II, indicating a preference for the American version of Roland II. The arrangement was that Hughes Aircraft would supply the firing units and the missiles would be bought direct from Aérospatiale.

US Roland is expected to attain Initial Operation Capability in the early 1980s.

Type: Surface-to-air, point defence
Configuration: Slender cylindrical body with cruciform delta wings, and four small fore planes near pointed nose. Tube launched
Length: 2.4 m
Diameter: 16 cm
Span (Max): 50 cm
Weight: 63 kg
Propulsion: Two-stage solid
Range: 500m-6.3 km
Guidance: Radio command. Optical aiming and automatic infra-red tracking of missile in Roland I. Roland II has radar target tracking instead of optical aiming
Warhead: High-explosive; proximity fuze
Main Contractors: Messerschmitt-Bölkow-Blohm GmbH, and Société Nationale Industrielle Aérospatiale (Euromissile). Hughes Aircraft/Boeing (US Roland)

Spada is an all-weather, short reaction time point/area missile system designed to defend airports, harbours, factories, bridges, rail junctions etc. from low and very low altitude attacking aircraft flying singly, in close formation or sequentially. Designed to employ semi-active homing missiles, and particularly the Aspide missile now in advanced development under an Italian Government contract, the system is modular in concept and its configuration can be adapted to optimise its deployment according to the nature of the surrounding terrain of the targets and of the threat.

A four-missile launcher is employed and other system elements are Tracking and Illumination Radars (TIR), Search and Interrogation Radars (SIR), and Control Unit. The simplest deployment includes one SIR, one or two TIRs, a number of launchers (up to four for each TIR) and a Control Unit, all co-located.

Both radars are of the coherent pulsed type. E/F-band for search, G/H-band for tracking and I-band for missile guidance.

A Mobile Spada system is also produced and includes: (1) a search and tactical control unit (with the capacity to control up to two fire control units); (2) a fire control and missile guidance unit (capable of controlling up to two launcher units); (3) a missile launcher unit.

All units are designed for installation on the FMC M548 tracked vehicle or on the FIAT TM 69 truck or any military vehicle with a load capacity of 5000 kg. The original mobility and transportability characteristics of the vehicles are in no manner degraded by the addition of the equipment shelters and launchers. The search and tactical control unit and the fire control and missile guidance unit are installed in self-contained shelters necessitating only a minimum of mechanical, inter-face with the carrying vehicles. The missile launcher unit comprises eight missile storage/launcher canisters mounted on a chassis which is trainable through 360°.

Design and development was undertaken by OTO-Melara and Selenia under Italian Government contract and deliveries are expected to start in 1978.

STINGER (XFIM-92A)

Formerly known as Redeye II, Stinger has been under development since 1972 as part of the US Army's Man-Portable Air Defence System (MANPADS) programme. It is intended as a more potent successor to Redeye. In announcing the project the Army said that Stinger would incorporate a high-performance propulsion system, improved infra-red devices and an advanced guidance technique which will give the weapon a greater range and velocity.

As so far developed, Stinger is a slightly larger weapon than Redeye: its overall length of some 152 cm is about 20 per cent greater than that of Redeye and it weighs about six per cent more — about 13.4 kg. Stinger also uses a passive infra-red homing, but its sensor works at a considerably shorter wavelength (4.1-4.4 microns) than Redeye and can thus home on the relatively high intensity exhaust plume of an aircraft engine.

In January 1974 the first guided test was carried out successfully, and the following year Stinger had six successful tests against manoeuvring targets between January and March. By December 1975 the contractor's (General Dynmamics) Engineering Development phase had been completed. In 1976 funds were allocated for the initiation of a new alternative type of seeker, a so-called 'two-colour' seeker, by which is meant infra-red and ultra-violet. This will give higher performance and greater resistance to counter-measures. If successful, about 75 per cent of future Stinger production rounds will be fitted with this seeker, the rest having the standard I/R unit. This programme should not be confused with Stinger Alternate, which is described separately.

STINGER ALTERNATE (USA)

Stinger Alternate is under development by Aeronutronic Ford Corporation and the US Army as a possible replacement for Stinger. The guidance technique employed is laser beam-riding, similar to that of the RBS 70. The same system is under consideration for use in anti-tank weapons also.

In Stinger Alternate, a shoulder-held guidance unit incorporates a stabilised sight-line for target tracking and a laser projector that is boresighted to the sight-line. The missile can be produced at low cost because it does not require a seeker head. The first guided flight of Stinger Alternate was made in July 1975, and the first fully successful guided flight test of an American supersonic laser beam-rider missile took place in June 1976 at White Sands Missile Range. Later in this series of tests a hit was scored on a QH-50D drone helicopter.

TIGERCAT

Tigercat is the land-based version of the successful Seacat surface-to-air missile, and is for low-level close-range air defence purposes. The basic fire-unit comprises a director trailer, a launcher trailer and two towing vehicles. These vehicles carry the firing unit crew of five men, generating equipment and miscellaneous support items. The launcher is a three-missile unit. Just as the shipborne Seacat is capable of accepting a variety of optional items of additional equipment. Tigercat can be operationally enhanced by means of alternative target acquisition and tracking equipment such as radar and TV.

Although Seacat was developed under UK Government contract, Tigercat was a private venture by Shorts. It has been in series production for several years and is in use, or planned for service, with forces of Iran, Argentina, India, Jordan, South Africa, and Qatar as well as the RAF Regiment.

Type: Land-based surface-to-air point defence system
Configuration: Uses Seacat missile with trailer-mounted launcher and guidance units
Length: 1.48 m
Diameter: 19 cm
Span (Max): 64 cm
Weight: Approx 63 kg
Propulsion: Two-stage solid
Range: 3-5 km Estimated
Guidance: Radio command. Optical or radar tracking
Warhead: High-explosive Contact and proximity fuzes
Main Contractor(s): Short Brothers and Harland Ltd

Tigercat is land version of the widely-used Seacat system

NAVAL MISSILE SYSTEMS

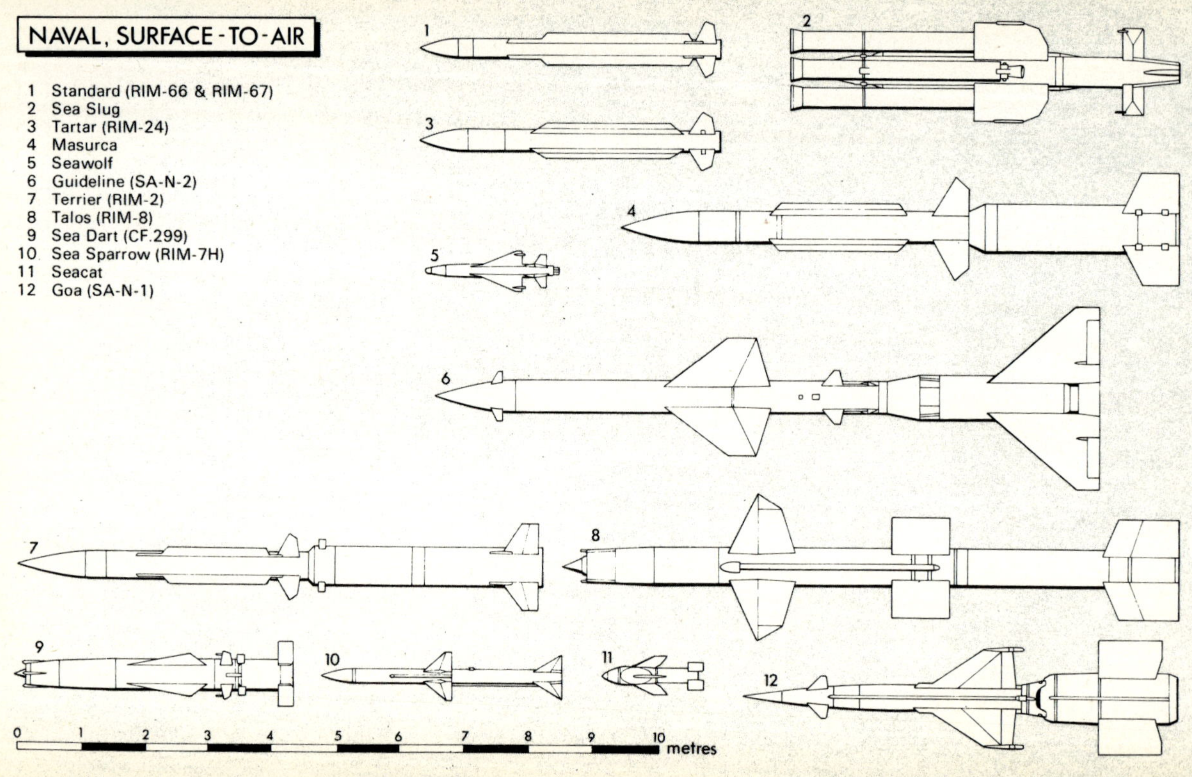
NAVAL, SURFACE - TO - AIR
1 Standard (RIM-66 & RIM-67)
2 Sea Slug
3 Tartar (RIM-24)
4 Masurca
5 Seawolf
6 Guideline (SA-N-2)
7 Terrier (RIM-2)
8 Talos (RIM-8)
9 Sea Dart (CF.299)
10 Sea Sparrow (RIM-7H)
11 Seacat
12 Goa (SA-N-1)
1
2
3
4
5
6
7
8
9
10
11
12
0 1 2 3 4 5 6 7 8 9 10 metres

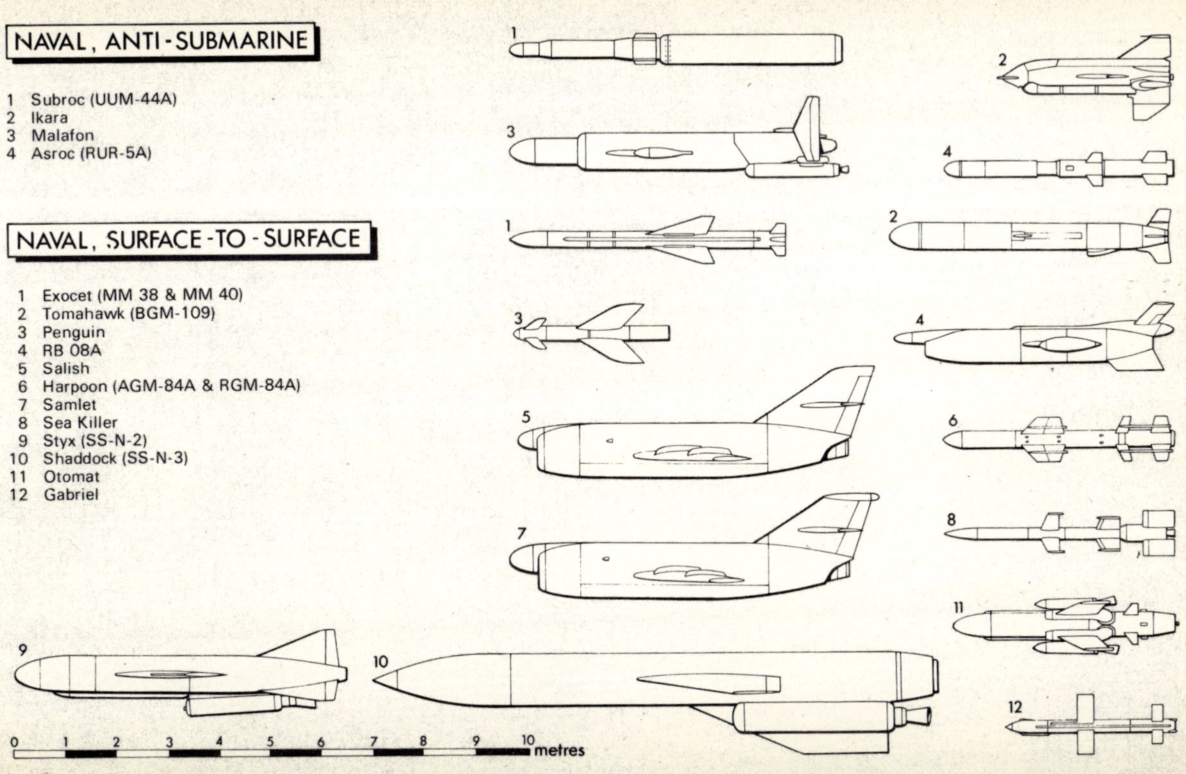

NAVAL, ANTI-SUBMARINE
1 Subroc (UUM-44A)
2 Ikara
3 Malafon
4 Asroc (RUR-5A)
NAVAL, SURFACE-TO-SURFACE
1 Exocet (MM 38 & MM 40)
2 Tomahawk (BGM-109)
3 Penguin
4 RB 08A
5 Salish
6 Harpoon (AGM-84A & RGM-84A)
7 Samlet
8 Sea Killer
9 Styx (SS-N-2)
10 Shaddock (SS-N-3)
11 Otomat
12 Gabriel
0 1 2 3 4 5 6 7 8 9 10 metres

NAVAL MISSILE SYSTEMS

This section of the book contains information on all naval missile systems with the exception of those included in other sections, such as air-launched weapons having a mainly naval role or Submarine-Launched Ballistic Missiles. In the Naval Missile Systems part of this volume will be found certain land-based weapons, for example Exocet and Otomat, which have the essentially naval function of coastal defence. Therefore the pages that comprise this section contain missiles of these various types, arranged in alphabetical order. However, it was considered more appropriate to arrange the scale drawings of them within the three main categories — Surface-to-surface, Surface-to-air, and Anti-Submarine Missiles. Certain of the missiles in the first two of these categories have alternative roles, and in such cases this is mentioned in the appropriate entry.

AEGIS

Aegis, formerly the Advanced Surface Missile System (ASMS), is the name of a sophisticated naval area coverage air defence system designed to protect USN formations from both aircraft and cruise missile attack. The missile to be used is a modified Standard SM-2 (described on a later page in this section), which is a semi-active radar homing weapon with provision for mid-course command guidance. It is launched from the Mk 26 fully automatic dual-purpose launcher which can also accommodate Asroc and Harpoon missiles. The launcher has a digital interface with the Mk 1 weapons control system which accepts weapon assignment commands and special threat criteria from the Mk 1 command and decision system and tracking data from the AN/SPY-1 multifunction radar. These inputs are processed to determine the possibility of engaging the target and then to generate commands for the Mk 26 launcher and pre-launch orders for the missile, commands for the Mk 99 fire control system for target illumination, commands to the multifunction radar, if mid-course guidance is required, and reports to the Mk 1 command and decision system.

The function of the Mk 99 fire control system is to illuminate, and if necessary track, the target. To do this it uses either the AN/SPG-62 (slaved) radar or the AN/SPG-63 (tracking) radar. Inputs to the system come from the Mk 1 weapons control system which, in the case of slaved radar operation is passing on data from the multifunction radar system.

The multifunction phased-array radar, AN/SPY-1 is a high-performance electronically-scanned equipment capable of surveillance and the simultaneous detection and tracking of multiple targets: it has been described as the heart of the Aegis system. Associated with it are four AN/UYK-7 computers.

Aegis planning started in 1964, and engineering development was entrusted to RCA in December 1969. By April 1972 the USN and RCA were able to announce completion of the system design phase. The single quadrant Aegis engineering development model completed successful land-based testing in 1973 and was installed in USS *Norton Sound* for sea trials in 1974. By early 1976, a number of improvements had been incorporated and 23 firings had been completed. In October 1976 RCA was awarded a $159m contract to develop a shipboard combat system using the Aegis fleet air defence system as its nucleus. It will incorporate a number of sensors and weapons in addition to Aegis. This contract runs until 1980.

Aegis weapon system structured test firing from the gunnery and missile test ship, USS Norton Sound. *(US Navy)*

ALBATROS

(ITALY)

Albatros is an all-weather missile and gun weapon system designed for short-range air defence against aircraft and anti-ship missiles. It also is capable of controlling both missiles and guns for surface engagements.

The complete weapon system includes the following major sub-systems:

(a) one or two missile and gun FCS
(b) one missile launching system
(c) up to three groups of guns
(d) Aspide missiles

Trials were carried out by the Italian Navy during 1973 off the Sardinia missile test range from the proving ship *Aviere* on which a prototype system was installed. The trials included launches against remote controlled air targets simulating anti-ship missiles; against surface targets, measurements of the system's ability to detect and counter air attacks under severe conditions; and checks of ECCM facilities. About 10 Albatros systems are in production for foreign frigates, while a contract for systems to be installed in second-generation frigates for the Italian Navy is under definition. Fittings in the Italian Navy will include four Lupo-class frigates, and ships of the same type of the Peruvian and Venezuelan Navies are also expected to fit the Albatros system. Four Greek destroyers will also be fitted.

Albatros firing from an Italian Navy ship

(INTERNATIONAL) ASMD

This is a US Navy development programme, in which there is now West German participation, directed towards a new shipboard weapon system for Anti-Ship Missile Defence. It is intended to supplement the anti-missile capabilities of other systems such as CIWS/Phalanx, the various point defence missile systems based on the Sparrow, and the longer-range capabilities of the US Fleet's other anti-air missiles. The prime threat which is to be countered is the cruise missile, and the overall ASMD weapons programme cites the Sea Sparrow (point defence) as the Ship's main battery in this role, with the ASMD Missile and the CIWS/Phalanx forming the secondary battery.

Four versions of the ASMD system are envisaged, the differing configurations depending upon whether or not the ship to be fitted (a) has a suitable digital Tactical Data System or Weapons Control Panel, and (b) if the ship has a Sea Sparrow launcher or not. Thus the four Types are:

(a) TDS with Sea Sparrow
(b) TDS without Sea Sparrow
(c) WCP with Sea Sparrow
(d) WCP without Sea Sparrow

The ASMD missile in its initial configuration consists of a 5in rolling airframe with dual-mode passive radio frequency midcourse and IR terminal homing. In this version the Redeye/Stinger IR seeker, the AIM-9L fuze (DSU-15B) and warhead (WDU-17B), and the Chaparral/Sidewinder propulsion motor are used. This guidance technique is target dependent in that it requires both active radar and IR signals from the attacking missile to function. However, a significant proportion of the Soviet inventory is of such a type and is thought likely to be so for some time to come.

In the near-term efforts, the US Navy's Passive Guided Projectile programme is also included in the ASMD effort. Both 5in and 8in laser guided projectiles using the Paveway technique are under development and a rocket motor is added to the shells to increase range and possibly speed also. An IR seeker is added to the guidance package to complement the semi-active laser seeker. The 8in guided projectile is expected to be ready for Fleet use aboard the first of the *Spruance*-class destroyers, and the 5in guided projectile in fiscal year 1982/83. For the more distant future (1985-90) the USN is embarking the Shipboard Intermediate Range Combat System (SIRCS) and Lightweight Modular Fire Control System programmes.

ASROC (RUR-5A) (USA)

Asroc is carried as the primary anti-submarine weapon aboard USN destroyers as well as some cruisers and frigates. It consists of a Honeywell Mk 46 acoustic homing torpedo or a nuclear depth charge attached to a rocket motor. It can be fired from an eight-cell launcher, the Mk 46 Launching System, or from the Mk 10 Terrier missile launcher. After launch, the weapon follows a ballistic trajectory, the rocket motor being jettisoned at a predetermined point. If the payload is a torpedo, it is lowered to the water by parachute, after which the homing mechanism is activated by immersion. Depth charge payloads are detonated at predetermined depths.

The first development contract was awarded to Honeywell in June 1956, after a two-month evaluation on the USS *Norfolk* in 1960, Asroc first entered service on four destroyers in 1961. It was fitted subsequently in eight Japanese destroyers, three West Germany destroyers, an Italian cruiser, seven Canadian destroyer escorts, and is now widely used in the US Navy.

Type: Ship-launched anti-submarine weapon
Configuration: Cylindrical body serving as a carrier rocket for either a homing torpedo (Mk 44 or Mk 46) or a nuclear depth charge. Cruciform tail fins and cruciform auxiliary fins set at 45 degrees
Length: 4.57 m
Diameter: 32.5 cm
Span (Max): 84.5 cm
Weights: Mk 44, 453 kg; Mk 46, 258 kg Approx
Propulsion: Solid
Range: 2-10 km
Guidance: Unguided in flight. Acoustic homing torpedo
Warhead: High-explosive MK 44 and MK 46 torpedoes, or nuclear depth charge
Main Contractor: Honeywell Inc

*Asroc anti-submarine weapon launch (*US Navy*)*

Having passed through successive stages of evolution, starting with the Murene/Mureca system proposals, the Naval Crotale as now adopted by the French and other, as yet unspecified, navies is an autonomous self-defence weapon system based on the successful land-mobile Crotale anti-aircraft missile. The naval weapon is intended for installation aboard high and medium tonnage ships for use against low-altitude air targets.

A major difference from the land-based Crotale is the 8-round launcher turret of the Naval Crotale, the container tubes of which are of rather larger diameter. The latter change is necessary to reduce the risk of interference with the missile's fins during launch due to ship movements. In other respects the land and shipboard versions of Crotale are very similar.

Naval Crotale has been selected for fitting in a number of ships of the French Navy, such as the *Georges Leygues* class, and sales of the system to other countries have been reported although these have not been identified. Series production continues. The present French Navy requirement is for 20 systems for fitting in 16 C70 corvettes, three F67 frigates and the PH75 Helicopter carrier.

Naval Crotale firing unit has a total of eight missiles ready to fire on 8-tube turret, the left-hand four being shown here

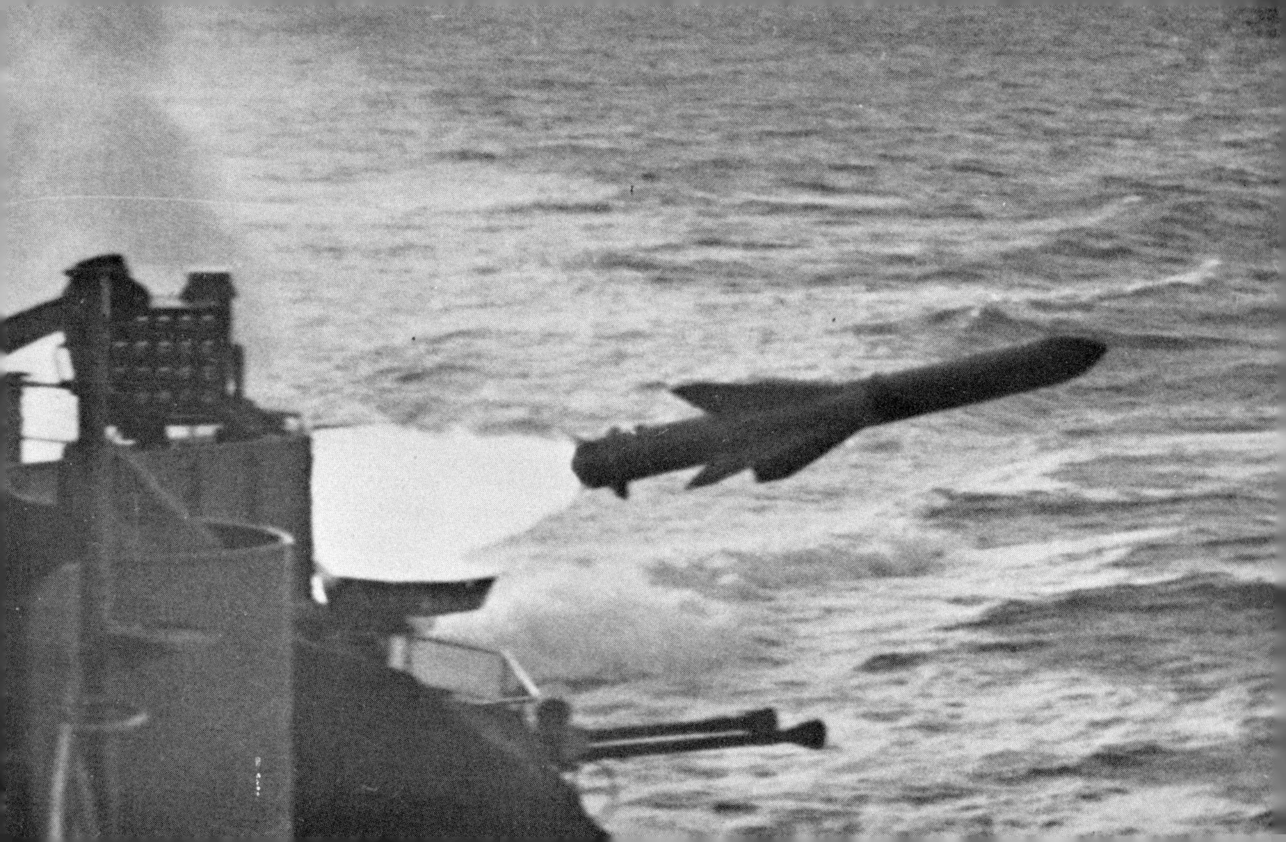

(FRANCE)

EXOCET (MM 38 & MM 40)

Exocet is a surface-to-surface tactical missile designed to provide surface warships with all-weather attack capability against other surface vessels. It can be fitted in major and minor warships including fast patrol boats and hydrofoils. It has also been developed for coastal defence purposes in a land-based role.

The missile flight consists of a preguidance phase during which it flies towards the target, whose range and bearing have been determined by the fire control computer and set up in the missile preguidance circuits before firing, and a final guidance phase during which the missile flies directly towards the target under the control of its active homing head. Throughout the whole flight the missile is maintained at very low altitude by a radio-altimeter.

The terminal guidance phase (approximately the last 10 km) is by means of the ADAC radar homing head developed by Electronique Marcel Dassault.

Subsequent to the introduction of the original MM 38 model, a number of derivative versions have been proposed, several of which have been produced. The air-launched AM 39 model is described in the section dealing with Air-Launched Missiles later in this book. The surface-launched variants are the result of combinations of: alternative propulsion motor configurations; new container/launchers of wound glass-fibre construction; and folding wings and tail surfaces. The basic warhead and guidance elements remain unchanged.

The MM 40 has a longer steel-cased sustainer motor than the MM 38, this resulting in a further increase in range. It is carried in a tubular glass-fibre container/launcher. Four of these launchers can be mounted in the space required for one steel MM 38 housing.

Essential differences between the three models of Exocet are illustrated in the following table.

Exocet model:	**MM 38**	**AM 39**	**MM 40**
Length:	5.21 m	4.69 m	5.64 m
Diameter:	35 cm	35 cm	35 cm
Maximum span:	100.4 cm	100.4 cm	100 cm
Weight:	735kg	650 kg	825 kg
Speed:	Mach 0.93		
Range:	42 km	50 km	70 km
Warhead:	165 kg	165 kg	165 kg
Guidance:	Inertial plus active radar homing		
Main Contractor:	Société Nationale Industrielle Aérospatiale		

Manufacturer's tests of the MM 38 were concluded in July 1972 and by the end of 1976, more than 1,000 Exocets and 150 ship installations had been ordered for a total of 16 navies. MM 40 development was continuing in early 1977 together with other Exocet development work. No details of coastal defence applications have been released. A submarine-launched version, SM 39, has been proposed.

Exocet anti-ship missile has a sea-skimming flight path to its target

FRAS-1

(USSR)

The designation FRAS-1 has been assigned to one of two types of anti-submarine missile believed to be associated with the forward twin launcher which first appeared aboard the Russian helicopter carriers *Moskva* and *Leningrad* in the late 1960s. The same launcher has since been observed on the first of the *Kiev*-class aircraft carriers, and is referred to as the SUW-N-1 launcher, indicating a surface-to-underwater function. The other weapon associated with this launcher is known as SS-N-14. No pictures of either type have been made public.

The FRAS-1 designation could be interpreted as Free Rocket, Anti-Submarine, One, and the weapon is thought to consist of a rocket with a range of some 30 km carrying a nuclear depth-charge warhead in a sumilar fashion to the American Asroc. The ship's own sonar and/or ASW helicopters would provide target location data for the FRAS-1 weapon system.

The second weapon, SS-N-14, is thought to be similar to the French Malafon or Australian/British Ikara in that a surface-to-surface missile carries a homing torpedo or a depth-charge payload to the vicinity of the target submarine. Estimated range is in the region of 37 km

Israeli Gabriel anti-ship missile is used by other navies also

GABRIEL

Gabriel is a shipborne anti-ship missile developed in Israel and designed for installation in ships from about 250 tons upwards.

A sea-skimmer, the missile is transported in, stored in, and launched from hermetically sealed, reinforced fibreglass containers. The standard launcher carries three missile cells on a rotatable pedestal, but single, fixed, cells also exist. The missile is powered by a solid-propellant rocket motor.

A sophisticated guidance and homing electronics system is carried by the missile. While details of this system have not been made public it appears that over the major part of its trajectory the missile is radar-guided in the horizontal plane and maintains its height above surface by means of a radio altimeter. A manual/optical alternative control system is available for use in conditions where radar guidance is not practical. When it nears the target the missile's homing system takes over: this has been described as a semi-active radar homing system, but it seems probable that an alternative form of homing is also available for use in ECM conditions.

Gabriel's cruising speed is in the high subsonic region. It was announced during 1972 that there were two versions of Gabriel in production, one having nearly twice the range of the other. It is believed, however, that the only differences between the two are that one has a larger propulsion unit and may incorporate additional guidance equipment for over-the -horizon operation.

Gabriel is now operational in fast patrol boats of the Israeli Navy, and is in service with five other countries at least, according to Israeli sources.

Sales and fittings of Gabriel are not widely publicised but it is no secret that the Israeli Navy operates the system aboard the *Saar*-class missile boats, and the *Reshef*-class boats. Other operators (some officially unconfirmed) are South Africa, Singapore, Argentina, and Malaysia.

In early 1976 it was reported that Israel has under development a successor to Gabriel Mk 2 in the form of a supersonic sea-skimmer anti-ship missile. Claimed to be suitable for both surface and air launch, it will have longer range and improved guidance. The provisional entry into service is forecast as the early or mid-1980s.

Type: Shipborne surface-to-surface
Configuration: Cylindrical body with pointed nose. Rectangular cruciform wings near mid-length, and similar, but smaller, tail surfaces
Length: 3.35 m
Diameter: 32.5 cm
Span (Max): 138 cm
Weight: 400 kg; Mk 2 is about 100 kg heavier
Propulsion: Two-stage solid. A more powerful motor is used in the Mk 2
Range: 22 km Long-range version reported with maximum of 41 km
Guidance: Autopilot plus radio altimeter cruise phase, possible with a command element from the launch vessel, and self-contained terminal homing
Warhead: High explosive 150 or 180 kg
Main Contractor: Israel Aircraft Industries Ltd

GOA (SA-N-1) (USSR)

Goa is the principal surface-to-air missile of the Soviet Navy. It is fitted to the 'Kanin', 'Kashin', 'Kotlin', 'Kresta' and 'Kynda' classes on scales ranging from one to four twin launchers. The shipborne Goa missile is assumed to be identical with the ground-to-air missile (SA-3) but the equipment associocated with it is very different. The standard launcher is clearly roll-stabilised as can be seen from the accompanying illustration. The launcher is mounted on top of the missile magazine and is reloaded vertically through small hatches. Associated with the missile is a compound radar system known by the NATO code-name Peel Group.

There is, of course no definite information on Goa's performance but it is generally taken to have maximum slant range of about 15 km and a ceiling of some 12,000 m. Length is estimated at about 5.9 m and the maximum span at 1.2 m.

Entry into service with the Soviet Fleet was in 1961-62. According to the latest available figures, Soviet vessels are equipped with a total of 128 launchers. There are two on each of eight SAM 'Kotlin' class, six 'Kanin' class, and four 'Kynda' class ships while 19 'Kashin' class, and four 'Kresta 1' class vessels each have four SA-N-1 launchers.

Rare picture of Soviet Goa, (SA-N-1) launch

GOBLET (SA-N-3)

The helicopter-carrier *Moskva* and Kresta-II class cruisers are equipped with a surface-to-air missile that is similar in size to Goa but possibly larger and certainly launched from a different launcher and directed by a different radar complex. Although there was confusion concerning this missile it now seems to be agreed that it is that classified in the American code as SA-N-3, its NATO code-name is Goblet.

The launcher does not appear to be roll-stabilised – suggesting that the missile gathering is efficient. The nature of the missile itself is unknown although various suggestions have been made. One suggestion, that appears to be consistent with the general Russian practice of making maximum use of any successful development, is that the missile is that known as Gainful (SA-6) and is a surface-to-air weapon of more modern design than Goa. Fire control radar has the NATO code Head Light. While the identification of Gainful with Goblet cannot be asserted, the suggestion has authoritative support.

Goblet-type launchers have been sighted on the two *Moskva* helicopter carriers, eight 'Kresta II' cruisers and on three 'Kara' class ships. In all cases, each ship had four launchers, to make a total of 52.

GUIDELINE (SA-N-2)

Guideline, also known, in its shipborne application, by the US alpha-numeric code SA-N-2, has been installed on only one Russian ship, so far as is known at present – the cruiser *Dzerjinski*. It has been suggested that one reason for this is the difficulty of providing suitable stable platforms on smaller vessels for the flapping Fan Song radars used to guide these missiles. It could also be that the relative difficulty that there appears to be in gathering these missiles to the required flight path presents even more of a problem in shipborne installations than it does on the ground.

So far as is known the general performance of Guideline as a shipborne missile is substantially the same as that described in the entry for the ground-based version.

HARPOON (AGM-84A & RGM-84A) (USA)

Air, surface ship, and submarine launched versions of the Harpoon tactical anti-ship missile have been developed. The designations AGM-84A and RGM-84A refer to the first two models, respectively, and the last is also known as Encapsulated Harpoon.

Harpoon is officially regarded as the principal American anti-ship weapon and has been designed for launch from all classes of USN surface ships (except patrol gunboats), the S-3 Viking and P-3 Orion maritime patrol aircraft, USN A-6 and A-7 attack aircraft, and in encapsulated form from all but the earliest nuclear attack submarines. Other nations that have, or are expected to place orders for either version of Harpoon are: Australia, Denmark, W. Germany, Iran, Israel, S. Korea, Netherlands, Norway, Saudi Arabia, Turkey, and the United Kingdom. The following description relates primarily to the ship-borne version, although much of the system is common to all three applications. The main body of the missile with its cruise-phase propulsion and guidance systems, homing and terminal manoeuvring systems, ECCM facilities, and warhead is common to all applications; the aerodynamic surfaces of this part of the system exist in several forms for compatibility with various aircraft, shipboard, and submarine launchers. All surfaces are designed for quick attachment and are interchangeable.

For all other than airborne launchings an additional boost section contains a solid-propellant boost motor. This propels the missile on a ballistic trajectory which it follows until the booster separates; after which the missile descends to a low cruise altitude, determined by its altimeter, and flies to the target under the power of its turbojet cruise engine. In the terminal phase the missile executes a 'pop-up' manoeuvre to evade close-in enemy defences and enhance the effectiveness of its warhead.

The weapon control system for Harpoon AN/SWG-I(V), is

produced by McDonnell-Douglas, with Sperry as subcontractor. Targeting data provided by shipboard systems is interfaced with the missile through this command and launch system. The Harpoon Data Processor digital computer receives targeting and altitude data from standard shipborne equipment (or from a third party for over-the-horizon operations) and computes the necessary missile and launcher orders. After launch, en-route guidance is provided by a missile-borne system consisting of a strap-down attitude reference assembly and a digital computer: no data inputs from the ship are required by the missile after it has been launched. Cruise altitude is monitored by a radar altimeter. Terminal guidance is achieved by means of a Texas Instruments DSQ-28 active radar homing system which maintains its lock until final impact. The radar homing system is frequency-agile; and this facility, coupled with extensive on-board computer logic circuitry, provides considerable ECCM capability to the missile in the terminal phase.

The weapon system development contract was awarded in July 1973 and entry into service began in 1977.

Type: Air-to-surface and surface-to-surface anti-ship missile. Submarine-launched version also in development
Configuration: Slender cylindrical body with pointed nose. Cropped delta cruciform wings and cruciform tail surfaces. Booster motor (when used) also has cruciform fins
Length: 3.84 m; 4.57 with booster
Diameter: 34 cm
Span (Max): 83 cm
Weight: 500 kg; 635 kg with booster
Propulsion: Solid booster. Teledyne CAE J402 turbojet cruise motor
Range: 110 km
Guidance: Programmed inertial plus radio altimeter cruise. Active radar terminal homing
Warhead: High-explosive 227 kg
Main Contractor: McDonnell Douglas Astronautics Company

Launch of Harpoon from hydrofoil, High Point. *(US Navy)*

IKARA

The concept of the Ikara anti-submarine weapon is the employment of a guided missile to deliver a homing torpedo to the target submarine. The missile is launched from a surface ship which uses a computer to calculate the torpedo dropping position from information fed into the computer from the ship's sonar and other sources of information on the submarine's position. The guidance system ensures that the missile flies to the continuously up-dated otpimum dropping position, derived from target information fed into the computer during flight. After release from the Ikara vehicle, the lightweight torpedo (such as the Mk 44) descends by parachute to the sea, where it searches for the target submarine and carries out a homing attack.

In the Australian version, the Ikara system has its own digital computer, whereas in RN systems the ships' Action Data Automation System (ADAS) provides this service.

More recently, the requirement to fit Ikara to the Brazilian *Niteroi* class Mk 10 frigates has given rise to the development of a third version of the system. Known as Branik, this version differs from the Australian and RN versions in the way in which the launcher and missile obtain computer service. In the *Niteroi'* weapon control system two fire-control computers (Ferranti FM 1600B) are used to control all the ship's weapons. The Branik system employs a special-purpose missile tracking and guidance system which is fully integrated with one of these computers. A lightweight semi-automated missile handling outfit is also incorporated in the new version.

A UK plan for an Improved Ikara having an extended range was suspended in 1976 for lack of funds.

Type: Ship launched anti-submarine
Configuration: Miniature aircraft configuration carrying homing torpedo
Length: 3.43 m
Span (Max): 1.53 m
Propulsion: Solid
Range: "To maximum sonar detection range"
Guidance: Command link. Acoustic homing for torpedo
Warhead: High-explosive
Main Contractors: Australian Department of Supply and Hawker Siddeley Dynamics Ltd

Australian Ikara anti-submarine weapon. System is used by Australian, Brazilian and British fleets

MALAFON (FRANCE)

Malafon is a shipborne ASW weapon consisting of a radio command guided winged vehicle carrying a homing acoustic torpedo. It is intended primarily for use from surface vessels against submarines, but may also be used to attack surface targets. Malafon has the appearance of a small conventional aircraft with short, unswept tapered wings, and a tailplane fitted with endplate fins. The missile is ramp launched and propelled by two solid-fuel boosters for the first few seconds of flight. Subsequent flight is unpowered. A radio altimeter is fitted to maintain a flat trajectory at low-level. On reaching the target area, approximately 800 m from the target's estimated position, a tail parachute is deployed to decelerate the missile. The homing torpedo is thus ejected from the remainder of the vehicle and enters the water to complete the terminal guidance phase of the attack by acoustic homing. Target detection and designation in the case of submerged targets is by means of sonar and by radar in the case of surface targets. These sensors, as appropriate, are used during the flight of the missile to provide data on the target for the generation of command guidance signals which are sent via radio command link to guide the missile. Missile tracking is aided by flares attached to the wing tips.

Development started in 1956 and by 1959 a total of 21 test launches had been made, 15 from the ground and six from an aircraft. The first sea launch and guidance test took place in 1962. Evaluation of the complete weapon system took place in 1964, during which time over 20 launches were made. Operational trials were carried out the following year. Deployment is limited to the French Navy, with installations on the two Suffren-class destroyers, three F67, one T56, five ASW T47, and one C65 ships.

Type: Shipborne anti-submarine. Some surface-to-surface capability
Configuration: Aircraft configuration with cylindrical fuselage housing acoustic homing torpedo in nose. Unswept tapered wings and tail plane with end-plate fins and rudders. Two jetisonable rocket motors for launch phase
Length: 6.15 m
Diameter: 65 cm
Span (Max): 3.3 m
Weight: 1 500 kg
Propulsion: Unpowered cruise phase. Two solid boosters for launch and acceleration phase
Range: 13 km
Guidance: Radio command with acoustic homing terminal phase
Warhead: High-explosive
Main Contractor: Société Industrielle d'Aviation Latecoere

MASURCA

Masurca is the surface-to-air missile system that forms the main anti-aircraft armament of the French destroyers *Duquesne* and *Suffren*, whose principal role is that of escort to the aircraft carriers *Foch* and *Clemenceau*. It has also armed the cruiser *Colbert* since 1973 when her modernisation was completed. Each of the Masurca ships is equipped with a DRBI 23 three-dimensional radar, a weapon direction system, two independent fire control systems using DRBR 51 radars, and a twin launcher. Two modes of guidance are employed and mixed salvoes using both techniques simultaneously are possible. The Masurca Mk 2 Mod 2 uses radio command, while the Mk 2 Mod 3 is equipped with a semi-active radar homing head. Externally the two versions are virtually identical.

Development of the French Navy's own long-range air defence missile system was the responsibility of the French Naval Arsenal at Ruelle, with appropriate assistance from specialist sectors of industry. In 1967 Engins Matra was asked to participate in the programme to bring its expertise to bear on the development of the radar homing version of the two types of missile used in this system.

Type: Surface-to-air. Shipborne anti-aircraft weapon
Configuration: Two-stage missile with booster of larger diameter than second stage. Latter is cylindrical with pointed nose and long-chord cruciform wings of narrow span. Cruciform tail surfaces on both stages.
Length: 5.29 m; with booster 8.6 m
Diameter: 41 cm
Span (Max): 1.5 m
Weight: Mod 2: 1 989 kg. Mod 3: 2 080 kg
Propulsion: Two-stage solid. Polka booster 34 000 kg. 2 080 kg sustainer in Mod 2, 2 170 kg sustainer in Mod 3
Range: 30-50 km. Approx
Guidance: Mod 2: Command/beam-rider. Mod 3: Semi-active radar homing
Warhead: High-explosive, 48 kg. Proximity fuze
Main Contractor: Direction Techniques des Constructions Navale

French-developed Masurca naval air-defence system

OTOMAT

Otomat is an anti-ship missile intended initially for launching from naval platforms of any size from fast patrol boat upwards, but also capable of land deployment (fixed or mobile), and use from aircraft. Weight for the aircraft-launched version is much reduced by elimination of the boosters which are not required in this mode of operation. The helicopter and land versions weigh the same as the shipborne model.

Target information can be derived from radar, active or passive search systems, or external data sources such as data link. The ability to gyro-angle the missile through 200° obviates the need for any ship manoeuvres at launch. After launching, the Otomat follows a cruise phase towards the target's predicted position, flying at low level under radio altimeter control and inertial guidance. An active homing head is used for the terminal phase.

A Mark II version of Otomat has been developed, the principal features of which are increased range (about 100 km) and the use of a new homing head produced by the Italian SMA concern (Segnalamento Marittimo e Aero). The latter allows Otomat to follow a sea-skimming flight path up to the point of impact with the target. It is reported that this version has provision for mid-course guidance updating. The Otomat Mk I, which is fitted with a Thomson-CSF Colvert homing head performs a late 'pull-up' and dive trajectory to the target. A third version, which has a reported range of 200 km, is in development for the Italian forces under the designation Teseo.

Development of the Otomat began in 1969 under a fixed-price contract and was jointly undertaken by Engins Matra in France and OTO Melara in Italy. The programme was first revealed in late 1970. Prior to the combined effort, each of the two concerns had been working individually in this area of missile technology. First test firings took place in early 1971, booster acceleration tests taking

place in Sardinia in April of that year. Guided test firings started in December 1971 and the first complete test of the system took place in February 1972 and was completely successful, the missile impacting on its target several tens of kilometres away. The first launch of Otomat Mk II took place in January 1974, and in June of the same year the Italian Navy hydrofoil, *Swordfish*, launched its first Otomat. In November 1975 the Italian Navy announced the successful operational firing of an Otomat from the recently commissioned Italian Navy hydrofoil *Sparviero* (Swordfish), of which six more have been ordered. The complete system is in full production and current orders are reported as numbering over 400 rounds. The system is being procured for the Italian, Peruvian, Venezuelan, and Libyan navies.

Type: Anti-ship missile, surface-to-surface and air-to-surface versions

Configuration: Cylindrical body with cruciform wings and tail surfaces. Air intakes at wing roots. Two jettisonable booster rockets on side of body

Length: 4.46 m

Diameter: 46 cm

Span (Max): 1.35 m

Weight: 770 kg

Propulsion: Arbizon III turbojet motor, 400 kg static thrust, plus two solid boosters for launch phase

Range: 60-80 km. Can be increased to 200 km at expense of payload

Guidance: Autopilot plus radio altimeter. Active radar terminal homing

Warhead: Semi-armour piercing high-explosive, 210 kg

Main Contractors: Engins Matra and Oto Melara SpA

Otomat anti-ship missile is seen here in coastal defence role

P 967

Penguin is a surface-launched anti-shipping missile primarily intended to provide small, fast naval craft with a powerful striking capability against larger vessels. It carries a 120kg Bullpup ASM-N-7A warhead at Mach 0.7 over a range of at least 20 km. The missile is mounted on a simple launcher which is built into a container which serves as protection and serves as packing for transport to and from the ship. Combined weight of box-launcher and missile is about 500 kg. Typical ship installations comprise four or six deck-mounted box-launchers.

Although originally designed for deployment on naval craft, the Penguin system is also being constructed for use in coastal defence and studies have been made of its use from helicopters. In the latter case, minor modifications only are stated to be necessary. Additionally, a modified version without the booster motor and with a smaller wingspan is being considered for use from jet fighters. The possibility of an active radar homing version, using a Swedish PEAB homing head, is also being examined.

Penguin employs inertial guidance for the cruise phase of its flight path, with passive infra-red homing to the target for the terminal phase. After launch the missile is thus independent of the parent vessel which is thereby freed to take evasive action or engage a further target. Target detection, acquisition and designation is by means of the launch vessel's radar and fire control system. Upon target acquisition, the fire control computer calculates the bearing so the predicted point of impact and the missile inertial guidance system is automatically slaved to the Kongsberg SM-3 computer data. The missile is then fired in the general direction of the target. After launch, the missile follows a programmed trajectory toward the predicted impact area. This programme can be varied. Prior to acquisition of the target by the missile's infra-red homing system this operates in a search mode, scanning a sector ahead of the flight path. Upon detection of the target, the homing system tracks the target and generates signals which are used to direct the missile to the target.

Operational in 'Snogg' class and 'Storm' class patrol boats and 'Oslo' class frigates of the Norwegian Navy, and on patrol boats of the Turkish Navy. An improved version (Mk 2) with longer range, will be installed on the Norwegian 'Hawk' class fast patrol boats now being built in Norway. Penguin has also been ordered for the Royal Swedish Navy.

Royal Norwegian Navy 'Storm' class FPB launching one of its Penguin anti-ship missiles

Type: Shipborne surface-to-surface anti-ship missile. Air-launched version in development
Configuration: Cylindrical body with tapered nose carrying cruciform swept-back fins. Swept cruciform wings set at mid-length. No tail fins
Length: 3 m
Diameter: 28 cm
Span (Max): 1.4 m
Weight: 330 kg
Propulsion: Two-stage solid
Range: 20 km
Guidance: Inertial plus infra-red terminal homing
Warhead: Semi-armour-piercing, Bullput ASM-N-7A warhead 120 kg. Impact fuze
Main Contractors: A/S Kongsberg Vaapenfabrikk

RB 08A

RB 08A is a surface-to-surface long-range cruise missile system intended for coastal defence and ship-to-ship use. Based on the Nord CT 20 target drone, the missile is a rocket-launched turbojet-powered monoplane carrying a warhead large enough to destroy an average freighter. The fire control system and launchers used vary according to the type of system (static coastal, mobile coastal, shipborne). The missiles are easily transported – the wings fold – and the amount of other equipment required for a mobile battery is not great. Range of the system is not disclosed, but it may be noted that the CT 20 drone has an endurance of 60 min and a maximum speed of 900 km/hr at 10,000 m. Since the all-up weight of the RB 08 is considerably greater than that of the CT 20 and since the RB 08 has the same power plant as that version of the CT 20 to which the figures above refer, a substantial reduction in endurance must be expected; nevertheless it seems probable that the system range is more likely to be limited by other factors than by the range capability of the missile.

The missile is launched by a booster unit which separates when expended. After climbing and further acceleration the missile reaches cruising altitude and speed; the climb being interrupted at a pre-set altitude which is then held by a constant altitude device. No details have been released concerning the operation of the homing equipment, but it seems likely that it is a radar operating in one of the higher frequency bands and capable of distinguishing between the target vessel and the surrounding sea clutter. The homing head is mounted in the nose of the missile; the warhead is mounted in the middle of the fuselage. It has been reported in France that homing heads for the RB 08A are supplied by Thomson-CSF.

Entry into service with the Swedish forces was in 1967 and production ceased in 1970.

Type: Surface-to-surface cruise missile. Ship-and land-launched versions
Configuration: Aircraft type layout with circular cross-section fuselage, tapered fore and aft. Swept wings at mid-length. Swept vee tail assembly. Radome protrudes ahead of nose air intake. Small ventral vertical tail fin
Length: 5.72 m
Diameter: 66 cm
Span (Max): 3.01 m
Weight: 900 kg
Propulsion: Turbomeca Mabore IID turbojet. Launcher powered by solid booster rockets
Guidance: Auto-pilot with self-contained homing for terminal phase
Warhead: High-explosive
Main Contractor: Saab-Scania AB

Swedish Navy uses RBO8A for coastal defence and ship-to-ship operations

SALISH

Salish is the NATO code-name given to a coastal defence cruise missile which is similar to and probably derived from the air-launched missile, Kennel, as also is the missile known as Samlet. Powered by a turbojet motor this missile has a range that has been estimated as anything from 100 to 200 km. Since Samlet appears to be the more modern of the two missiles it may be wise to assume not more than 100 km for Salish. When shown publicly, Salish has appeared mounted on a short ramp on a low trailer. Mounted for launching, however, it has been seen on a larger ramp and with a JATO bottle attached.

Like Samlet, Salish has a radome above the air intake in the nose. This is believed to cover a semi-active homing equipment. There are no other obvious indications of guidance equipment. Approximate dimensions are given below in the entry for Samlet.

SAMLET (USSR)

Like Salish, the missile known to NATO as Samlet is a surface-to-surface version of the air-to-surface jet-powered cruise missile whose NATO code name is Kennel. It has been operational in the USSR, Poland, Egypt and Cuba as a coastal defence weapon. When shown publicly Samlet has been mounted on what appears to be its launching ramp constructed as a trailer. For launching a JATO bottle is presumably necessary.

Samlet, like Salish, has a radome over its jet air intake. The radome is larger than that on Salish and Samlet also has what seems to be an electronics pod mounted on its tail fin. The radome is assumed to cover a radar homing seeker: the other device may contain the receiver and associated apparatus of a command guidance link. Apart from these differences the two missiles appear very much alike and the following approximate data are probably equally true of both.

Length: Approximately 7.0 m
Wing span: Approximately 5.0 m
Weight: Believed about 3 000 kg
Cruising speed: Mach 0.8-0.9
Range (Samlet): Possibly as much as 200 km with mid-course guidance

SA-N-4

This designation (SA-N-4) has been allocated to whatever missile system is concealed within the 'bin-type' launcher assemblies which first appeared on the Nanuchka missile boats and Krivak destroyers and spread fairly rapidly to a number of other classes of Soviet Navy ship. A retracting mechanism is involved so that a 'pop-up' missile launching mode is possible. A twin launcher is employed. Most reports assign a close-in air defence role to the SA-N-4, with some observers stressing the anti-helicopter aspect. Each SA-N-4 'bin' is associated with a fire control radar group, code-named 'Pop Group'. As for the missile itself and its possible configuration and mode of guidance, totally inadequate evidence upon which to embark on any worthwhile conjecture is so far available.

However, the disclosure in the late 1975 of a previously unknown land-mobile air defence weapon, the SA-8, during a Red Square military parade, gave more fruitful grounds for speculation. Gecko, as the SA-8 is known to NATO, was shown with its vehicle-mounted radar group and missiles exposed for all to see. Readers are referred to the Gecko entry on a preceding page for what details there are. All that we feel justified in adding here is that there are apparent similarities of some consequence in the Pop Group radars of the SA-N-4 and those of the SA-8; the missile of the latter would fit the SA-N-4 launcher arrangements so far as it can be judged; and the operational performance requirements of the respective land-based and naval systems are sufficiently close to make the proposition of a broadly common system an attractive one.

The Soviet Fleet has SA-N-4 launcher installations on six classes of vessel. Two converted Sverdlov cruisers have one each, the new 'Kara' class of cruisers have two amidships; five 'Krivak' class ships each have two; six Nanuchka missile boats have one each; and 11 'Grisha I' class corvettes have a similar foredeck installation to the Nanuchkas. The new Kiev-class aircraft carriers have two SA-N-4 installations.

SCRUBBER (SS-N-1) (USSR)

The SS-N-1 cruise missile is the earliest known weapon of this kind deployed by the Soviet Fleet. Two code-names have been associated with the SS-N-1, Strela and Scrubber, but on the evidence available, the latter is regarded as the correct NATO designation. Little is known with any certainty of the precise dimensions or configuration of Scrubber beyond the fact that it has an aircraft-type layout and requires a launch ramp about 17 m in length. This ramp is carried on a mount that can be trained over an angle of about 200° and elevated. It is surmounted at one end by a 'hangar-type' enclosure, presumably used to house one missile in a 'ready-to-fire' state. To the rear of the launcher a deckhouse is located for the housing and preparation of reload missiles.

The SS-N-1 Scrubber is thought to fly at subsonic speeds and to carry a high-explosive warhead. The probable means of propulsion is by ramjet, with a solid-propellant booster rocket for the initial launch and acceleration phases. A maximum range of about 185 km has been credited to Scrubber in some quarters, but the practical operating range is probably considerably less. Infra-red homing during the terminal phase is considered likely, although opinions on this differ and it is possible that alternative homing heads may have been developed during the long life of this weapon. Entry into service was in 1958-59 with 'Krupny' and 'Kildin' class destroyers of the Soviet Fleet. The numbers of these vessels fitted have decreased over the years as a result of conversions, and by 1975 only one installation, in a Krupny class destroyer, was known to remain. It can therefore be presumed to have been withdrawn by now.

SEACAT

Development work on Seacat started in the 1950s and the first sea trials took place on HMS *Decoy* in 1962. Since then this missile has been adopted by at least fifteen of the world's navies, including the RN, and has also been adapted for land-based operations as the Tigercat. The initial launcher for naval use was a four-round equipment but a lightweight three-round launcher was a later development and now both types are in widespread use. Similarly the fire control arrangements for Seacat have undergone steady modification and improvement. The Royal Navy's optical director system is the GWS 20, while the radar versions are the GWS 21 and GWS 24. Other navies equipped with Seacat use both optical and radar directors of a variety of nationalities. The Swedish designation for Seacat is the RB 07 and in this case it is used with a fire control radar of Dutch origin.

Type: Surface-to-air, shipborne missile. Surface-to-surface capability. Land-based version also
Configuration: Generally cylindrical body but with fore portion having flattened sides towards the pointed nose. Cruciform swept wings and cruciform square profile tail surfaces
Length: 1.48 m
Diameter: 19 cm
Span (Max): 64 cm
Weight: Approx. 68 kg
Propulsion: Two stage solid
Range: 3.5 km Estimated
Guidance: Radio command. Optical or radar tracking
Warhead: High-explosive. Contact and proximity fuzes
Main Contractor: Short Brothers and Harland Ltd.

Seacat serves in at least 15 of the world's navies

SEA DART (CF.299) (UK)

Sea Dart is a third-generation naval area defence weapon system with the ability to engage high and low-altitude aircraft targets as well as certain types of missiles. Performance data are classified but excellent range is understood to be a feature and the ramjet engine enables the missile to fly at controlled speeds throughout its flight envelope. There are fully automatic magazine handling and loading arrangements for the twin launcher.

Development was started in August 1962 and test firings began in 1965. The first production contract was announced in November 1967. Development work was completed by the early 1970s and the first system entered service in HMS *Bristol,* to be followed by the Type 42 destroyers of the Armada Republica Argentina and the Royal Navy. The RN designation is GWS 30.

Type: Surface-to-air, with surface-to-surface and anti-missile capability
Configuration: Cylindrical body with larger diameter booster motor at rear. Nose intake for ramjet sustainer has four interferometer aerials on its rim. Broad stub cruciform wings and small tail fins on missile. Four large fins on booster section
Length: 4.36 m
Diameter: 42 cm
Span (Max): 91 cm
Weight: 550 kg
Propulsion: Solid booster: liquid fuel ramjet sustainer
Range: At least 30 km
Guidance: Semi-active radar
Warhead: Presumed high-explosive
Main Contractor: Hawker Siddeley Dynamics Ltd.

Sea Dart missiles on twin launch ramp

SEA KILLER Mk 1

Sea Killer Mk 1 (at one time also known as Nettuno) is a short-range shipborne surface-to-surface missile. Development was initiated by Contraves Italiana SpA in 1963 and the first prototype was completed in 1966. The first operational installation was made in 1969, in which year responsibility for the missile system was taken over by Sistel Spa. The Sea Killer Mk 1 is in service with the Italian Navy.

Type: Shipborne surface-to-surface
Configuration: Slender cylindrical body with pointed nose. Cruciform wings at mid-length and rectangular cruciform tail surfaces
Length: 3.73 m
Diameter: 20.6 cm
Span (Max): 85.7 cm
Weight: 170 kg
Propulsion: Solid, 2 000 kg static thrust
Range: 10 km plus
Guidance: Beam-rider/radio command/radio altimeter, with optical/radio command reversionary mode for use under interference conditions
Warhead: High-explosive, 35 kg, fragmentation with impact/proximity fuze
Main Contractor: Sistel SpA

Sea Killer Mk1 firing

SEA KILLER Mk 2

(ITALY)

Similar in many respects to the earlier Sea Killer Mk 1, the Mk 2 Sea Killer (once bearing the Italian designation Vulcano) is a shipborne surface-to-surface tactical missile. It is larger than Sea Killer Mk 1 and is powered by a two-stage rocket motor. Sea Killer Mk 2 can be used in the Marte helicopter-launched anti-ship missile system. Development of Sea Killer Mk 2 was initiated by Contraves Italiana SpA in 1965 and the first prototype was completed in 1969. In that year the Sistel organisation was formed and responsibility was transferred from Contraves. The Sea Killer Mk 2 system is operational on four Vosper-Thornycroft Mk 5 frigates of the Imperial Iranian Navy.

The Sea Killer Mk 2 missile is also used in the air-launched anti-ship weapon system Marte, which is described in a later section of this book. A version of the Marte system for use aboard small naval craft has been developed under the name Mariner.

Type: Shipborne surface-to-surface. Also used in Marte air-launched anti-ship system
Configuration: Slender cylindrical body with pointed nose and with booster motor at rear end. Second stage has cruciform wings at mid-length and cruciform tail surfaces. Booster stage has large rectangular cruciform stabilisation surfaces
Length: 4.7 m
Diameter: 20.6 cm
Span (Max): 100 cm
Weight: 300 kg
Propulsion: Two-stage solid. SEP 299 booster 4 400 kg static thrust. SEP 300 100 kg static thrust sustainer
Range: 20 km
Guidance: Beam rider/radio command/radio altimeter, with optical/radio command mode for use under interference condtions
Warhead: High explosive semi-armour-piercing. Impact/proximity fuzing
Main Contractor: Sistel SpA

Sea Kíller Mk2 launch

SEASLUG

Development of the Seaslug began in the early 1950s with prototype trials being carried out in HMS *Girdleness* in the latter part of that decade. The only ships equipped were the eight RN County Class GMDs. Of these the first four were fitted with the MK 1 Seaslug while the later ships of the class received the more powerful MK 2. The first, HMS *Hampshire*, has since been paid off, leaving seven in the class. First fittings of Seaslug took place in 1961.

Type: Surface-to-air. Shipborne. Some surface-to-surface capability
Configuration: Cylindrical body with pointed nose. Cruciform wings and tail surfaces. Four strap-on, jettisonable boosters attached to fore body at launch
Length: 6.1 m
Diameter: 41 cm
Span (Max): 1.44 m
Weight: Classified
Propulsion: Four solid boosters. Solid sustainer
Range: Better than 45 km. Estimated
Guidance: Radar beam-rider
Warhead: High-explosive, proximity fuze
Main Contractor: Hawker Siddeley Dynamics Ltd

Seaslug naval air-defence missile at the moment of launch. Royal Navy's County-class GMDs were only ships fitted

SEA SPARROW (RIM-7H) (USA)

Sea Sparrow is the general name given to a number of applications of the Sparrow air-to-air missile to naval surface-to-air applications. In addition to the United States Navy, those of Belgium, Canada, Denmark, Italy, Japan, Netherlands, Norway, and possibly Spain have all utilised this basic missile in a variety of shipboard air defence systems. The USN systems are the Basic Point Defence Missile System (BPDMS) and the Improved PDMS; the Canadians have their own Canadian Sea Sparrow system; the Italian Albatros system (already mentioned on a previous page) can use Sparrow as an alternative to Aspide; and there is also the NATO Sea Sparrow system. The main differences lie in the launcher and sensor arrangements employed.

The Basic Point Defence Missile System was assembled on an urgent basis in the mid-1960s from existing hardware. It features the Sparrow III (AIM-7E) missile launched from a modified eight-tube ASROC missile launcher. The launcher is housed on a modified 3in, 50 calibre automatic gun carriage. Total weight of the system is 17,690 kg. A CW semi-active radar homing system is used.

The improved PDMS incorporates a new lightweight eight-cell launcher, a fire control system using digital computers, a Target Acquisition System (TAS) featuring a dual-mode sensor and a powered director/illuminator. The Sparrow missile family will still be used but in a modified form (RIM-7H). The version of the missile has folding fins which enable it to fit into a smaller launcher. The improved system will be produced as a co-operative effort by the US and five other NATO countries – Belgium, Denmark, Italy, Norway, and the Netherlands, this being known as NATO Sea Sparrow. The US is expected to purchase approximately one-half of all systems produced.

The Canadian Sea Sparrow uses the AIM-7E2 version of the missile, carried on a four-round pylon mounted on an extendible cantilever beam. The launcher head is remotely rotated in azimuth and elevation. A Dutch Signaal M22/6 fire control radar system serves both missiles and guns.

the improved system will be produced as a co-operative effort by the US and five other NATO countries – Belgium, Denmark, Italy, Norway, and the Netherlands, this being known as NATO Sea Sparrow. The US is expected to purchase approximately one-half of all systems produced.

The Canadian Sea Sparrow uses the AIM-7E2 version of the missile, carried on a four-round pylon mounted on an extendible cantilever beam. The launcher head is remotely rotated in azimuth and elevation. A Dutch Signaal M22/6 fire control radar system serves both missiles and guns.

Sparrow AIM-7E as it leaves Sea Sparrow Launcher. (US Navy)

SEAWOLF

Seawolf is the Royal Navy's short-range self-defence missile system, GWS25. Designed to provide rapid reaction defence against both aircraft and anti-ship missiles, it is capable of installation in new and existing small escort vessels down to about 2,000 tons, full load, as well as in larger vessels. Lightweight derivatives of the GWS25 system, known as Seawolf/Psi, Seawolf/Omega and Seawolf/Delta, have been studied for fitting in much smaller vessels of corvette size or possibly as little as 400 tons. Seawolf/Psi is a blind-fire variant, Seawolf/Omega is a visually directed variant and Seawolf/Delta is a lightweight darkfire version.

The Seawolf missile employs line-of-sight guidance with radar differential tracking or television, both with radio command. Speed and manoeuvrability characteristics are suitable for the engagement of small Mach 2 missile and aircraft targets under severe weather conditions and sea states.

Co-mounted Type 967 and Type 968 radars provide air and surface surveillance facilities and the tracker radar is the Type 910. Integral IFF facilities are included. The Vickers Mk 25 Mod O six-round launcher is standard with the GWS25, but the company has designed lightweight twin and three-rounds launchers for use in the alternative Seawolf systems.

First studies began in 1964 and on their completion in 1967 BAC was nominated as the missile contractor, with Marconi responsible for radars and the overall ship system. Full development started on July 1, 1969. Prior to that, BAC had proposed and carried out six successful firings of a longer range version in 1969, but this was not pursued. Land tests of the adopted Seawolf at Woomera included hits on supersonic Petrel rockets and even 4.5 inch shells. The sea trials aboard HMS *Penelope* that followed yielded similarly favourable results. Production started in 1976 and the system is due for fitting in the RN's new Type 22 B-class anti-submarine frigates, and it will replace the Seacat/GWS22 systems on some Leander-class frigates. Some Type 21 frigates may also be fitted. American firms working on the USN Shipboard Intermediate Range Combat System (SIRCS) programme have signed an agreement with BAC covering the provision of Seawolf technology.

Type: Surface-to-air. Shipborne weapon with anti-missile capability
Configuration: Cylindrical body with conical nose. Cruciform delta shaped wings and tail surfaces. Container-launched
Length: Classified, about 2 m
Diameter: Classified
Span (Max): Classified
Weight: Classified
Propulsion: Solid
Range: Classified
Guidance: Beam-rider/command with options of radar, optical, and TV tracking
Warhead: High-explosive
Main Contractor: British Aircraft Corporation

SHADDOCK (SS-N-3)

Too little has been seen of the SS-N-3 Shaddock cruise missile on public display for it to be described with confidence. The general view seems to be that it is a ramjet or turbojet-powered cruise missile with hinged wings that open out when the missile leaves its cylindrical launcher. Initial boost is provided by two rocket units under the rear of the fuselage. Length of the missile is estimated at about 10 m and the diameter of the fuselage is probably around 1.0 m.

Command guidance is used; and for surface vessel installations the missile is tracked by Scoop Pair radar and course corrections transmitted to it by radio. Russian technical literature has described methods of terrain following for cruise missiles, but without referring to any specific missile but it is quite probable that a measure of terrain following capability is provided for Shaddock in its overland role. This relies upon a radio altimeter which would also be useful in the anti-ship uses of Shaddock. For the terminal phase it is believed that infra-red homing is used; but it may be assumed with some confidence that this is one feature of the missile that will have been affected by modifications in the 10 years or so that these missiles have been in service, the most likely development being the provision of an active radar homing head.

Missile speed is believed to be transonic and range is limited mainly by radio/radar horizons. With mid-course guidance the maximum achievable range is believed to be about 550 km, but practical ranges are nearer to 180 km for missiles launched from cruisers and less than that for submarine-launched missiles. It is believed that the standard warhead is nuclear with a yield in the kiloton range.

Shaddock-type missiles are installed operationally in 'Kresta I' and 'Kynda' class guided missile cruisers and in submarine classes E1, E2, J and W. The E1 and E2 nuclear submarine classes carry respectively six and eight missiles in pairs, the launchers being let into the hull so as to present a smooth surface except when elevated for launching. The 'J' class non-nuclear submarines have similar launching arrangements for four missiles. In the 'W' class submarines two arrangements are still current – one being the Long Bin arrangement described above and the other the 'Twin Cylinder' arrangement in which two launching tubes are mounted on the deck aft of the conning tower.

On surface ships both fixed and trainable multiple container mountings are used and there is a truck-mounted version for coastal defence and/or surface tactical missions. Successor for Shaddock is believed to be the SS-N-12.

Shaddock anti-ship cruise missile launcher for the coastal defence role of this weapon. It is also carried on ships

Type: Surface-to-surface. Ship- and submarine-launched, land-based version also
Configuration: Believed to be of basic aeroplane configuration with stub wings and two solid booster motors which are jettisoned after burnout
Length: 10.0 m Provisional
Propulsion: Two solid boosters for launch and acceleration phase. Ramjet or turbojet cruise motor
Range: Estimates vary widely. Maximum practical range probably 320 km
Guidance: Probably auto-pilot plus radio altimeter with radio command for cruise. Active radar and/or infra-red terminal homing
Warhead: Probably nuclear, Kiloton range

SHIELD (UK)

Shield is the name given to a feasibility study of a naval close-range air defence system based on the Hawker Siddeley Dynamics SRAAM (Short Range Air-to-Air Missile). The missile would incorporate an infra-red seeker homing head but of increased sensitivity to that used in the air-to-air role, otherwise the two versions would be the same. Greater seeker sensitivity would enable lock-on to be achieved against targets at all aspects, including head-on. Quick reaction is stated to be another feature of the system but no details have been disclosed of specific target search and acquisition systems or equipment considered. Neither have any particular launcher configurations been specified, although the possibility of adapting a number of existing lightweight launchers has been examined. Shield is proposed for use as the main anti-aircraft weapon in small craft or as a close-in weapon system in larger ships. The study was carried out by Hawker Siddeley Dynamics under British Ministry of Defence contract.

Shield is the designation for the projected use of the SRAAM as the weapon for a naval close-in air-defence system

SKA

At the 1975 Paris Air Show it was revealed that the Saab-Scania Aerospace Division has been working for some time on three new-generation missile projects, of which one is a possible new surface-to-surface missile to serve as an alternative armament for the Swedish Navy torpedo boats and flotilla leaders. Other possible applications for this weapon, designated SKA, include fixed or mobile coast artillery roles, and as an air-to-surface missile for the Swedish Air Force. (The other two new missile projects are a new version of the Saab RB 05, and the Saab 372 air-to-air missile, both described in the Air-Launched Missile section of this volume).

The SKA is of the 'launch and leave' type and will have an effective homing head with both sea and ground target tracking capability. It is stated to be highly resistant to countermeasures and to combine relatively small weight and dimensions with excellent long-range performance. No other reliable information has been obtained at this stage. A Swedish Government decision (based on field tests and studies) on possible full-scale development and production was expected in 1977.

SLAM

(UK)

The initials SLAM stand for Submarine (or Surface) Launched Air Missile system. This system has been developed by Vickers to meet the need of submarines and light surface craft for an effective short-range defence against other surface craft and helicopters.

For target engagement the system uses the Blowpipe missile for which a special multiple launcher is provideed. This carries six missiles clustered around a central electronics enclosure which contains part of the missile control equipment, television system and gyro subsystem for launcher stabilisation. Target acquisition is by means of attack periscope, the launcher being automatically aligned with the target in azimuth when the launcher mast is raised. The operator then seeks the target's elevation and tracks it on his TV screen, controlling the launcher system by means of a thumb button controller which enables him to maintain the target in the screen centre. He then selects and fires the missile. When the missile is fired, the thumb button controller is disconnected from the launcher control circuits. The missile is automatically gathered on the line of sight and appears on the TV screen, at which point the operator controls its flight with the same thumb button controller. Missile range is at least 3,000 m and may be greater against slow moving or stationary targets because less energy is required for manoeuvring. The warhead weighs 2.2 kg and is detonated by impact or proximity fuzes.

Details of fittings have not been released but Israel and Brazil are both known to have British-built submarines and to have shown interest in SLAM.

SS-N-7

Little is known of the SS-N-7 beyond the fact that it is a member of the Russian family of cruise missiles developed in recent years. It is carried by C-class Soviet nuclear submarines on a scale of eight launchers per submarine. Launching from a submerged vessel is possible. The missile is reported to have autopilot cruise control and some form of homing system. Range has been quoted as between 45 km and 55 km and it has been suggested that a major portion of the flight is carried out in a sea-skimming mode.

SS-N-9

The designation SS-N-9 has been allocated to the surface-to-surface missiles carried in the two triple launcher/containers aboard the Soviet 'Nanuchka' class missile boats which made their appearance in 1969. To date, no pictures or official details of the missiles themselves have been made public, and hence performance figures must be regarded as provisional. No other class of vessel has been definitely associated with the SS-N-9, and neither has any NATO code-name for this weapon been made public. It has been conjectured, however, that the anticipated new 'Papa' class of Soviet submarine may employ SS-N-9 or a derivative as its cruise missile armament, although the existing SS-N-7 may be retained for this class of boat.

The estimated range of the SS-N-9 is up to about 275 km with external mid-course guidance by co-operating aircraft or helicopter. A normal operating range of about 75 km seems likely. Autopilot, with or without radio command link guidance, is the probable method of cruise phase control and active radar homing may be the normal terminal homing technique. The associated search and fire control radar group on 'Nanuchka' class vessels is reported to be code-named Band Stand.

Fittings so far confirmed are confined to the 'Nanuchka' class of missile boats, each of which has two triple launchers to give a total of 84 launchers at the end of 1976. The possibility of Nanuchkas being purchased by the Indian Navy has been reported, but not confirmed.

SS-N-10 (USSR)

The designation SS-N-10 has been allocated to the surface-to-surface missiles carried in new-pattern container/launcher fitted aboard 'Kara', 'Kresta II', and 'Krivak' class vessels of the Soviet Navy. As yet no NATO code-name for this weapon has been made public. Neither has any official data or any photograph of the missile itself, and therefore any published performance figures must be regarded as provisional. The SS-N-10 is thought to have a cruising speed in excess of Mach 1 and a range of about 55 km has been quoted. It is thought likely that the figure of 55 km might apply to the maximum autonomous range, ie. when launched without the assistance of a co-operative aircraft for mid-term cruise guidance control. No reliable information as to the form of terminal homing used has been obtained, but the most likely candidate is active radar, possibly with an optional anti-radiation passive homing mode.

A more recent possibility which has been advanced in Western defence circles is the existence of an SS-N-14 which is believed to be a surface-to-surface weapon with a range of about 37 km and armed with an ASW torpedo. In addition to forming part of the armament of Soviet helicopter carriers, it has been suggested that the SS-N-14 may be an alternative round for the SS-N-10 launchers. The general concept of the SS-N-14 is alleged to be similar to the French Malafon system.

The SS-N-10 became operational in 1968, and the latest available deployment figures show a possible total of 140 launchers. These are distributed as follows: eight 'Kresta II' with eight launchers; 11 'Krivaks' each with four; and four 'Kara' class light cruisers with eight launchers each. More ships of all three classes are under construction.

SS-N-11

The designation SS-N-11 has been given to the surface-to-surface missile carried in the new type container/launchers seen on the latest version of 'Osa' missile boats of the Soviet Fleet. The missiles themselves have not been seen in public but are generally believed to be an advanced version of the SS-N-2 Styx. The performance is assumed to be similar in terms of range and speed, but the guidance techniques are expected to be more advanced. In addition to the 55 'Osa II' class vessels with which the SS-N-11 was initially associated it has recently been learned that a modified version of the Soviet Navy 'Kildin' class of destroyer has four SS-N-11-type launcher containers, two on each side of the after funnel, both pointing aft. It is possible that these have been installed as a replacement for the elderly SS-N-1 Scrubber at one time seen on this class of destroyer. Two twin gun turrets have been fitted in the space vacated by the large Scrubber launch assembly.

Two modified Kashin class destroyers have since been seen with four SS-N-11 aft-facing launchers. This gives a total of at least 244 SS-N-11 launchers in service.

SS-N-12

The cruise missile associated with the new pattern launchers that made their appearance aboard the new Soviet aircraft carrier *Kiev* in late 1976 have been designated SS-N-12. They are thought to be an improved version of the Shaddock, SS-N-3, or a replacement for it. Little is known of the missile or its performance and it has not yet been revealed publicly. The *Kiev* has two quadruple mountings forward for SS-N-12.

SS-N-15

The designation SS-N-15 applies to an anti-submarine missile which is understood to be similar in principle and operation to the American Subroc (UUM-44A). A range of about 45 km has been reported and a nuclear payload is possible. Victor-class submarines are reported to be armed with the SS-N-15 but it seems probable that others may also carry a mixture of missiles and torpedoes for ASW.

SS-NX-13 (USSR)

The SS-NX-13, according to official US sources, is a tactical ballistic anti-ship missile which may have been intended for deployment aboard Russian Y-class nuclear submarines. When this book was printed, the SS-NX-13 had not been tested since November 1973 and had not entered operational service. However, it was stated that the advanced technology employed was significant and the project could be revived.

STANDARD (RIM-66 & RIM-67)

Development of the first Standard missiles began in December 1964, the first operational models being the shipborne RIM-66 and RIM-67 Standard MR and ER (Medium-Range and Extended-Range) respectively. The function of these two missiles was to replace progressively the Tartar and Terrier missiles used for air defence by USN frigates. Since then the basic Standard has been applied to a number of additional roles, some of which are the subject of current development activity. It is now operational in two versions with a third in devleopment. First is the Semi-active Standard (SM-1-MR); second, Standard ARM (Anti-Radiation Missile), which is also for air launch; and third Active Standard, which has an active radar homing head for over-the-horizon operations. Yet another version is the Standard (SM-2) which is being developed for the Aegis fleet defence system, and which will combine mid-course guidance facilities and terminal homing. The SM-2 will be deployed in both MR and ER versions, as with SStandard SM-1, but the newer models will have typical ranges which are much greater, 48 km and 96 km, respectively. SM-2 will be used by the USN to augment or replace Tartar, Terrier and Standard SM-1 missiles.

The various models and their US designations are listed below:

AGM-78: Standard ARM (Anti-Radiation Missile) for air launch. Similar to RGM-66D

YRIM-66A: Standard MR basic development model

RIM-66B: Similar to YRIM-66A but with improved motor, Mk 56 Mod O

RIM-66C: Standard MR similar to RIM-66B but adapted for Aegis system

Standard SM-2 at sea for firing trials in 1977. This version will be used in USN Aegis system

RGM-66D: Standard ARM, surface-to-surface version
RTM-66D: Training version of RGM-66D
RGM-66E: Similar to RGM-66D but adapted for use with Asroc launcher
RIM-67A: Standard ER. Uses Mk 30 Mod 1 booster to give extended range

It is also US practice to refer to those models of Standard not intended for use with Aegis as Standard SM-1, while the later Aegis variants are known as Standard SM-2. The SM-1 continues to undergo improvement and recently a new warhead, Mk 90, and fuze, Mk 45, were incorporated.

Type: Surface-to-air shipborne missile in medium-range (MR) and extended-range (ER) versions. Variants of both versions are used in other systems (Aegis, Standard ARM) for other shipborne and air-launched roles.

Configuration: Cylindrical body with pointed nose and cruciform long-chord narrow-span wings. Cruciform tail surfaces. ER has tandem booster with cruciform tail surfaces.

	MR	**ER**
Length:	4.57 m	8.23 m
Diameter:	30.5 cm	30.5 cm
Weight:	590 kg	1,360 kg
Propulsion:	Dual-thrust solid	Two stage solid
Range:	24 km	56 km

Guidance: Semi-active radar homing. Active homing, and passive anti-radiation homing heads developed for other applications of Standard
Warhead: High-explosive. Direct action or proximity fuze
Main Contractor: General Dynamics Corporation

STYX (SS-N-2)

The SS-N-2 Styx anti-ship missile is a short/medium range weapon which is very widely deployed by the Soviet Fleet and the navies of many of its allies. Entry into service was in 1959 or 1960, but despite the length of service this indicates, new Styx installations are apparently still being made.

The general configuration is that of a small aircraft, with a delta planform wing and a triple tail-surface arrangement. A jettisonable booster rocket is used for the launch and acceleration phases, after which an internal motor sustains a cruising speed of about Mach 0.9. Range is estimated as about 40 km maximum.

During the relatively long service life of Styx it is probable that several guidance modes have been employed, and some references quote the existence of 'A' and 'B' versions of the SS-N-2 (but without agreement as to the differences between the two models). They probably relate to various combinations of the following alternatives: (1) the cruise phase could be carried out under either autopilot or radio command guidance; (2) the terminal phase could rely upon continuation of command guidance, active radar, or infra-red homing. Over the years, the launcher/hangars associated with Styx have undergone progressive changes and it seems reasonable to assume that the guidance apparatus also has received periodic updating, in which case all of the above methods may be employed in different installations.

A total of about 1,100 Styx missiles is operational aboard vessels of the Soviet and at least 16 other navies. The People's Republic of China is an important user of SS-N-2 and there is good evidence that after having been supplied with initial quantities of Styx missiles and 'Komar' and 'Osa' missile boats, the Chinese have now established their own production lines. In addition, China seems to have been the first nation to add Styx to the armament of larger vessels. The new class of destroyers of indigenous design, but based on the Soviet 'Kotlin' class in some respects, has two triple launchers for what is almost certainly Styx (Chinese version) amidships. The same source has also reported that China has a number of Styx missiles deployed as coastal defence weapons.

Soviet picture of Styx in a land-based role does not show the booster rocket employed in the normal shipboard installations

SUBROC (UUM-44A)

(USA)

Subroc forms the offensive element of an advanced US Navy anti-submarine weapon system designed for use in nuclear-powered attack submarines operating against submarines armed with strategic missiles. The missile is launched horizontally from a standard 21-inch (53.3 cm) torpedo tube, by conventional means. At a safe distance the solid-fuel motor is ignited and Subroc follows a short level path before turning upward to leave the water for the major portion of its trajectory. At a predetermined point, separation of the nuclear depth bomb is initiated by explosive bolts and a thrust reversal deceleration system which enables the warhead to continue on a ballistic trajectory controlled by vanes on the depth bomb. At a preset depth, the nuclear charge is detonated. Each attack submarine carries four or six Subroc missiles.

Subroc development began in 1958 at the US Navy Ordnance Laboratory with Goodyear Aerospace Corporation as prime contractor. Technical evaluation was completed in 1964 and production and operational deployment began in 1965. Production is expected to continue to 1978 at least.

Type: Submarine-launched, air-flight, anti-submarine weapon
Configuration: Cylindrical body with fore-part of smaller diameter than first stage. Second stage has four tail fins and short ogival nose
Length: 6.4 m
Diameter: 53 cm
Weight: 1 815 kg
Propulsion: Two-stage solid
Range: Up to 56 km
Guidance: Inertial
Warhead: Nuclear depth charge
Main Contractor: Goodyear Aerospace Corporation

Subroc warhead balancing (above), *and the Talos long-range naval air-defence missile* (right)

TALOS (RIM-8)

The Talos series of shipborne missiles is one of the most powerful in the US Navy inventory. Its main function is in the anti-aircraft role for fleet defence, but later versions are also capable of surface-to-surface operation. US military designations run from RIM-8A to RIM-8J.

The basic Talos missile is a beam-rider with a semi-active homing terminal phase. Propulsion is by solid booster and a ramjet sustainer motor. The latter burns a mixture of kerosene and naphtha, and total thrust is quoted as 9,070 kg.

Of the more recent variants in operational use the RIM-8G for high-altitude interception at long range employs beam riding mid-course guidance, with semi-active CW interferometer terminal guidance. The RIM-8J is similar to the RIM-8G but with an improved homing system. One model is provided with an anti-radiation homing head. This is the version intended for surface-to-surface missions and is consequently now designated RGM-8H. Both HE and nuclear warheads are available with Talos missiles and there is provision for the remote selection of either type from the magazine for automatic loading on to the launcher.

Talos will probably eventually be replaced by the Aegis missile system, but initially it will be the Tartar and Terrier missiles which will be replaced and Talos replacements will not start until Aegis has a longer-range missile than the SM-2: current plans call for Talos to remain operational until the early 1980s.

Type: Surface-to-air and surface-to-surface shipborne missile
Configuration: Cylindrical body tapering slightly to nose air intake containing centre-body of conical shape. Interferometer antennas round edge of intake. Cruciform wings and tail surfaces. Cruciform tail fins on booster motor
Length: 6.4 m, 9.53 m with booster
Diameter: 76 cm
Span (Max): 2.9 m
Weight: 3175 kg
Propulsion: Solid booster, liquid fuel ramjet sustainer
Range: 120 km
Guidance: Beam-rider with semi-active radar homing in basic version. RIM-8H has anti-radiation homing
Warhead: Alternative high-explosive or nuclear
Main Contractor: Bendix Corporation

TARTAR (RIM-24)

(USA)

Tartar is a medium-range missile and it provides primary air defence for US Navy destroyers and destroyer-escorts, and secondary air defence for cruisers. It is no longer in production but in addition to the USN, this weapon is in the inventories of a number of foreign navies including those of France, Italy, Japan and West Germany. Entry into American service was in 1961, and it is progressively being replaced in the USN by the Standard.

Type: Surface-to-air. Shipborne
Configuration: Cylindrical body with pointed nose. Long-chord, narrow-span cruciform wings, and small cruciform tail surfaces
Length: 4.57 m
Diameter: 34 cm
Weight: 646 kg
Propulsion: Dual-thrust solid
Range: 16 km
Guidance: Semi-active radar homing
Warhead: High-explosive. Direct action and proximity fuzes
Main Contractor: General Dynamics Corporation

Tartar air-defence missile on Mk 13 Mod 3 launcher

TERRIER (RIM-2)

The Terrier series of anti-aircraft missiles has been in operational service since 1956. It has also been the subject of sustained development and improvement but it is now progressively giving way to the Standard missile in US Navy service. Most versions of Terrier are armed with a conventional HE warhead, but one model, the RIM-2D, was produced with a nuclear payload. Development was started in 1951 and entry into operational service followed about five years later. The Advanced Terrier (RIM-2F) became operational in 1963. In addition to the USN the Italian and Netherlands Navies also are equipped to operate Terrier.

Type: Surface-to-air. Shipborne Surface-to-surface capability
Configuration: Cylindrical body with pointed nose and long-chord, narrow-span cruciform wings. Cruciform tail surfaces. Booster of larger diameter has its own cruciform tail surfaces
Length: 8.23 m with booster
Diameter: 34 cm; Booster 46 cm
Weight: 1360 kg
Propulsion: Two-stage solid
Range: 35 km. Estimated
Guidance: Beam-rider plus semi-active radar homing
Warhead: High-explosive. Proximity fuze
Main Contractor: General Dynamics Corporation

*Terrier launch from a US Navy Ship. (*US Navy*)*

TOMAHAWK (BGM-109) (USA)

A tactical version of the US Navy Tomahawk is planned in addition to the strategic model which is described in the appropriate section of this book. Nuclear and conventional warheads will be available for both versions and the principal difference will be in the range of about 550 km for the tactical model.

AIR LAUNCHED MISSILES

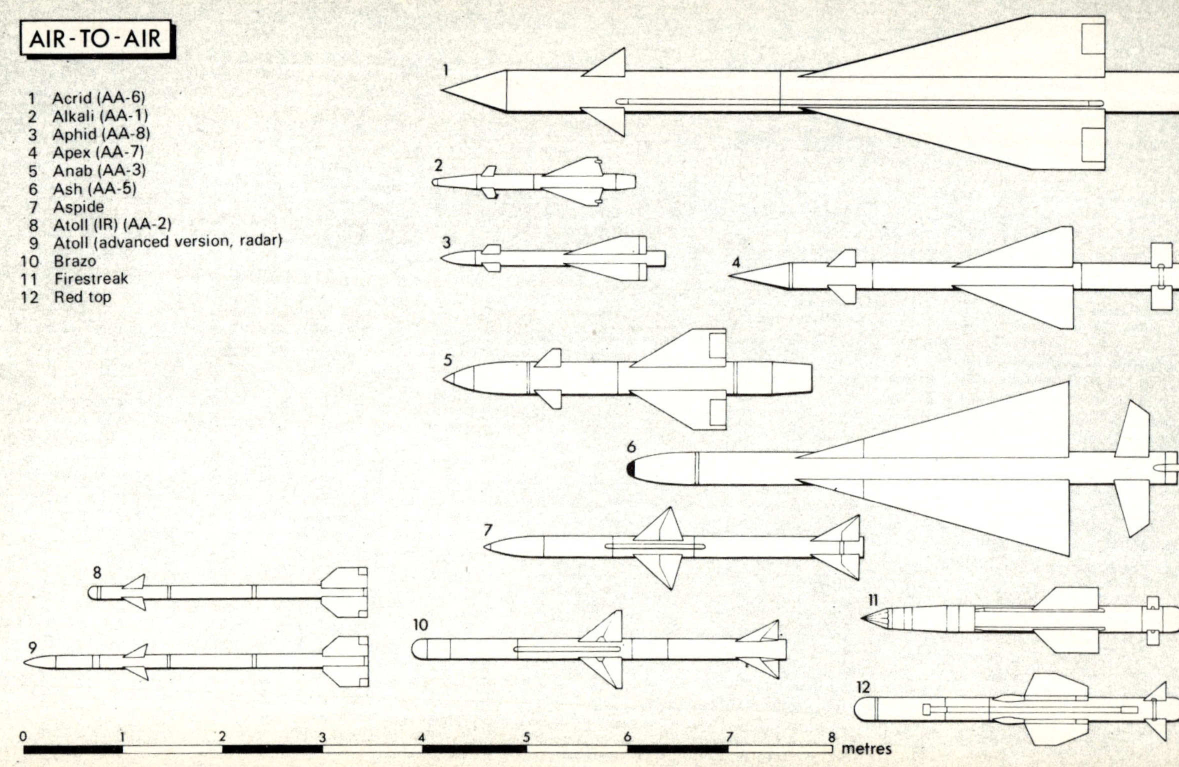
AIR-TO-AIR
1 Acrid (AA-6)
2 Alkali (AA-1)
3 Aphid (AA-8)
4 Apex (AA-7)
5 Anab (AA-3)
6 Ash (AA-5)
7 Aspide
8 Atoll (IR) (AA-2)
9 Atoll (advanced version, radar)
10 Brazo
11 Firestreak
12 Red top
1
2
3
4
5
6
7
8
9
10
11
12
0 1 2 3 4 5 6 7 8 metres

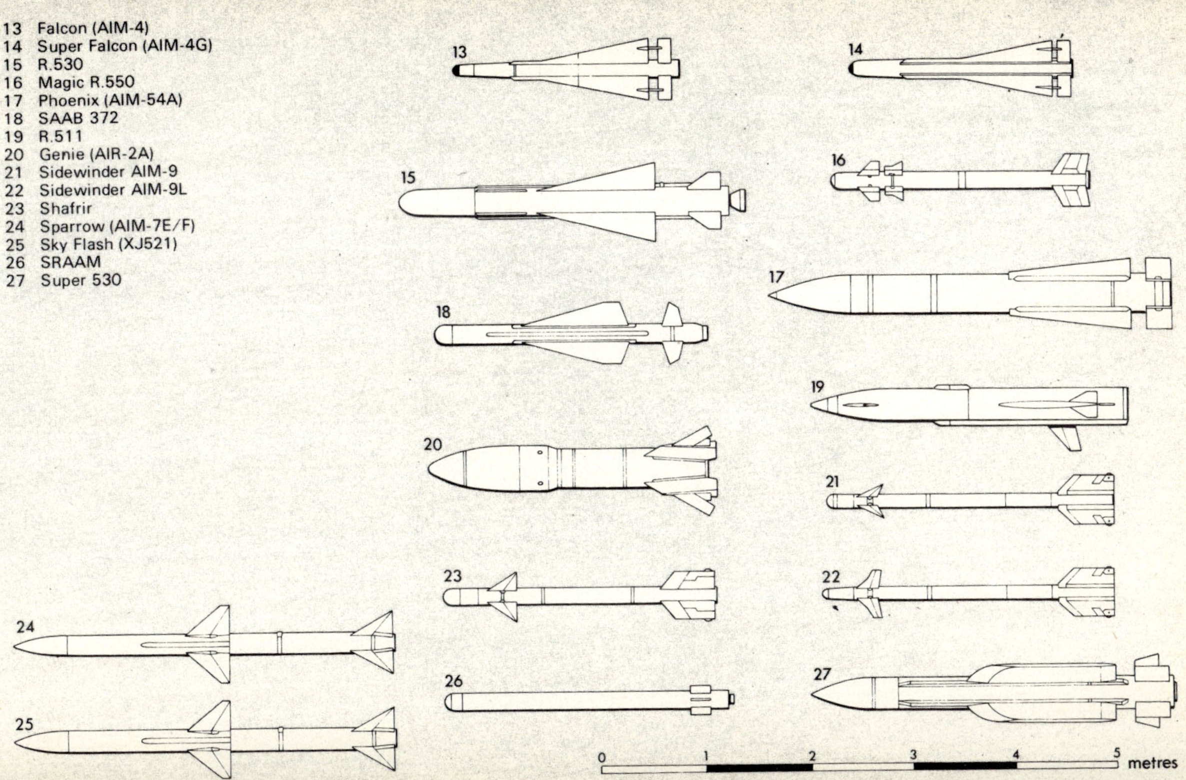

13 Falcon (AIM-4)
14 Super Falcon (AIM-4G)
15 R.530
16 Magic R.550
17 Phoenix (AIM-54A)
18 SAAB 372
19 R.511
20 Genie (AIR-2A)
21 Sidewinder AIM-9
22 Sidewinder AIM-9L
23 Shafrir
24 Sparrow (AIM-7E/F)
25 Sky Flash (XJ521)
26 SRAAM
27 Super 530
13
14
15
16
17
18
19
20
21
23
22
24
26
27
25
0
1
2
3
4
5
metres

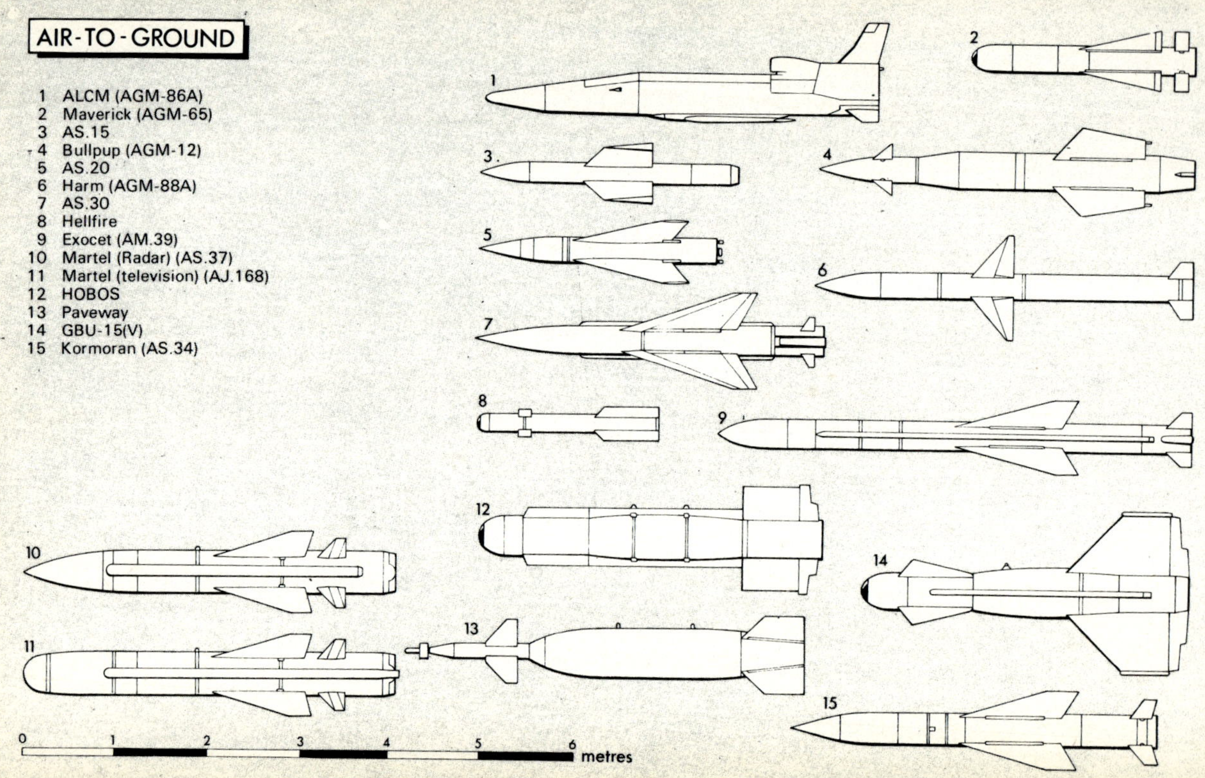
AIR-TO-GROUND
1 ALCM (AGM-86A)
2 Maverick (AGM-65)
3 AS.15
4 Bullpup (AGM-12)
5 AS.20
6 Harm (AGM-88A)
7 AS.30
8 Hellfire
9 Exocet (AM.39)
10 Martel (Radar) (AS.37)
11 Martel (television) (AJ.168)
12 HOBOS
13 Paveway
14 GBU-15(V)
15 Kormoran (AS.34)
1
2
3.
4
5
6
7
8
9
10
11
12
13
14
15
0
1
2
3
4
5
6
metres

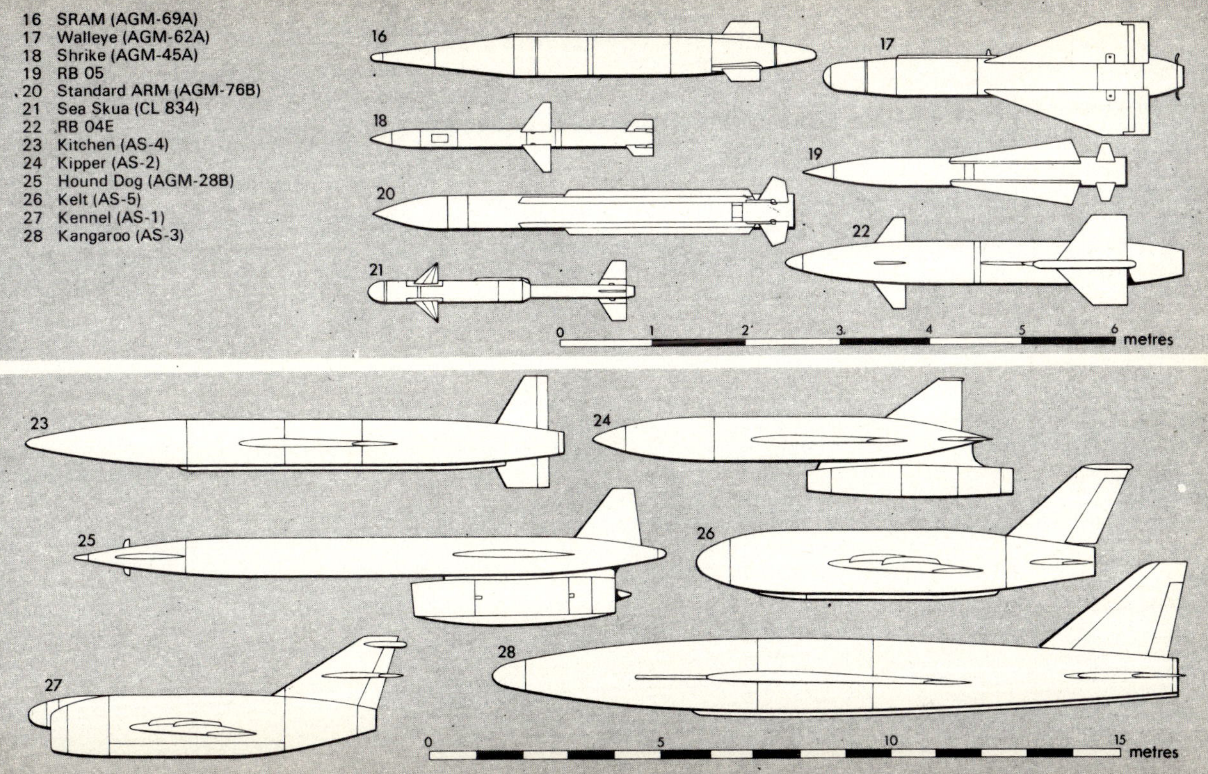
16 SRAM (AGM-69A)
17 Walleye (AGM-62A)
18 Shrike (AGM-45A)
19 RB 05
20 Standard ARM (AGM-76B)
21 Sea Skua (CL 834)
22 RB 04E
23 Kitchen (AS-4)
24 Kipper (AS-2)
25 Hound Dog (AGM-28B)
26 Kelt (AS-5)
27 Kennel (AS-1)
28 Kangaroo (AS-3)
16
17
18
19
20
21
22
0 1 2 3 4 5 6 metres
23
24
25
26
27
28
0 5 10 15 metres

AIR-LAUNCHED MISSILES

This section deals with two main categories of air-launched weapons: air-to-air missiles and air-to-surface missiles. Both classes are listed together in one alphabetical sequence, but in the scale drawings each of the two groups is illustrated separately.

The air-to-air missiles all have the same basic role of air-defence or as a means of asserting air-superiority over a battle area. The functions of the air-to-surface weapons are more varied, and range from strategic missions, through differing tactical applications on land and sea, to straightforward ground attack or battlefield support roles. Airborne applications of anti-tank missiles, such as HOT and TOW, are not covered in the following section and the reader is referred to the Anti-Tank Missiles section of the book.

Advancing technology in air-to-surface weapon development has necessitated the inclusion of a number of items which by some definitions are not missiles, on the grounds that they lack their own propulsion system. Such weapons, also known as PGM (Precision Guided Munitions), warrant their inclusion by virtue of the fact that various means of guidance after launch from the carrying aircraft are provided. Walleye and the GBU-15 are examples of this class of weapon, which is one of growing military significance.

AAM-1

The AAM-1 was developed by Mitsubishi Heavy Industries Ltd as a replacement for the Sidewinder on F-86 and F-104J interceptor aircraft of the Japanese Air Self-Defence Force. No official details have been released but there is reason to accept the generally published information that the AAM-1 is an infra-red homing weapon with conventional HE warhead, a likely range of about 7 km, and weighing about 70 kg at launch. This last characteristic and a reported missile length of 2.60m support the view that the AAM-1 may well be based on a development of the Sidewinder. At the time of entry into service, late 1970, it was variously reported that a total procurement of between 330 and 3,000 missiles was planned. The lower figure is more likely in view of the emergence of an AAM-2 model having collision course interception capability, thus improving on the AAM-1's pursuit-course capability. It was stated in 1976 that a total planned production of 330 had been completed by late 1971.

AAM-2

The AAM-2 development programme was started by Mitsubishi Heavy Industries Ltd as a more advanced successor to the AAM-1 air-to-air missile in 1971-72. No official details have been released, although one important difference between the AAM-2 and its predecessor is that it has a collision course interception capability. Infra-red homing guidance and a conventional high explosive warhead are generally presumed, but there is no reason to rule out the possibility of radar homing, or even alternative homing heads of both types for the AAM-2.

The planned procurement for Fiscal Year 1976 was 40 prototype rounds to be used in air trials due for completion by late 1977. Favourable results and an appropriate Government decision could lead to production deliveries in 1979.

ACRID (AA-6) (USSR)

Probably the largest air-to-air missile in the world today, the Acrid, AA-6, is standard armament on the Soviet MiG-25 Foxbat interceptor, which is capable of carrying four on under-wing pylons. Two versions have been observed, equipped with radar and infra-red homing heads, respectively. Most of the little that is publicly known about Acrid is based upon deductions from a single picture of Foxbat carrying four such missiles, and which was first published in September 1975. The first set of estimated dimensions for Acrid were derived from this photograph and the provisional dimensions of the Foxbat aircraft. When the defecting Russian pilot, Lt. Belenko, delivered a MiG-25 to Japan in September 1976, accurate measurements of the aircraft were possible, which in turn enabled fresh estimates to be made of Acrid's proportions. These proved to be somewhat different from the original estimates but can be assumed to be the more accurate and are therefore used in the following table.

Type: Air-to-air
Configuration: Cylindrical body with rear-mounted long-chord wings and small canard control fins. Nose profile differs between radar and IR versions
Length: 6.29 m (radar), 5.91 (IR)
Diameter: 36 cm
Span (Max): Aprox 1.76 m
Weight: 700-800 kg Estimated
Propulsion: Solid
Range: About 50 km (radar), 20 km (IR)
Guidance: Probably semi-active radar, and IR
Warhead: HE about 60 to 90 kg

AA-6 Acrid, thought to be the world's largest air-to-air missile

Full load of AA-1 Alkali air-to-air missiles. The tail cone is jettisoned when the rocket motor ignites

Alkali is the NATO code-name assigned to the Soviet air-to-air missile normally deployed on the MiG-19 interceptor, and also reported as having been carried by the MiG-17 and the Su-9. Estimated length is 188 cm, diameter 18 cm. The US designation is AA-1. Span of the rear fins is about 58 cm, and of the forward set 32 cm. Use of a solid-fuel motor is assumed. Range is probably not more than 6-8 km, and speed between Mach 1 and 2.

Alkali guidance is generally believed to be radar based in view of the deployment of this missile on all-weather interceptors, in which case it is probably of the passive homing type, with the launch aircraft fire control radar providing target illumination.

Type: Air-to-air
Configuration: Cylindrical body, tapered at both ends. Four delta shaped wings aft and four small delta control surfaces forward
Length: 1.88 m Provisional
Diameter: 18 cm Provisional
Span (Max): 58 cm Provisional
Propulsion: Solid
Range: 6-8 km Provisional
Guidance: Probably semi-active radar
Warhead: High-explosive

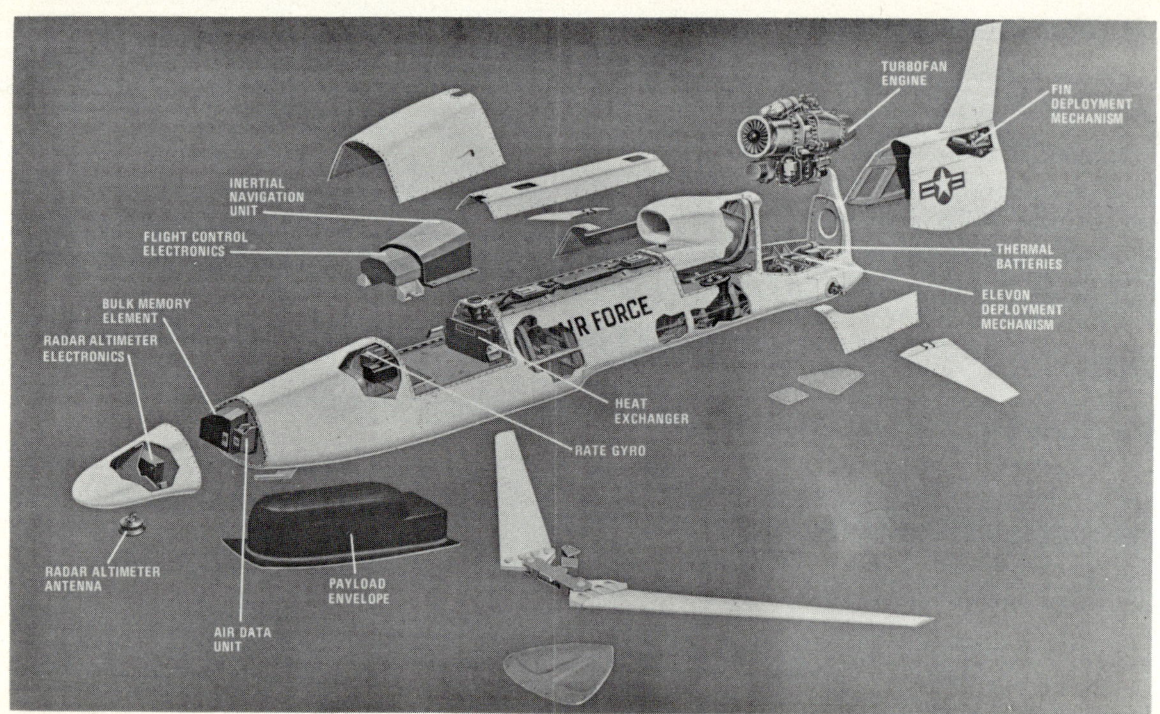
TURBOFAN ENGINE
FIN DEPLOYMENT MECHANISM
INERTIAL NAVIGATION UNIT
FLIGHT CONTROL ELECTRONICS
THERMAL BATTERIES
ELEVON DEPLOYMENT MECHANISM
BULK MEMORY ELEMENT
RADAR ALTIMETER ELECTRONICS
AIR FORCE
HEAT EXCHANGER
RATE GYRO
RADAR ALTIMETER ANTENNA
PAYLOAD ENVELOPE
AIR DATA UNIT

ALCM (AGM-86A)

The Air-Launched Cruise Missile, AGM-86A, programme is intended to develop a weapon for use by the USAF strategic bomber force. The initial carrier will be the B-52 but ALCM will be used on the B-1 and possibly the FB-111 also. The ALCM programme will make maximum use of the now-cancelled SCAD (Subsonic Cruise Armed Decoy) engineering development programme for vehicle design and turbofan engine development. This design is particularly suited for launching from strategic bombers in that it can complement the SRAM on a one-for-one basis in both internal rotary racks and pylon mounts.

The general configuration of the ALCM can be seen from the illustration. For internal stowage, all flying and control surfaces are folded or retracted, being deployed automatically after release from the aircraft. Dimensions in flying configuration are: length 4.26 m, span 2.89 m, and weight about 900 kg. The engine is Williams Research Corporation F107-WR-100 turbofan developing about 272 kg of thrust for a weight of 60 kg. Operating speed is described as 'high sub-sonic'. The design is intended for high-speed cruise flight at very low altitudes for distances of up to 1,300 km, and to aid penetration of oposing defences a very low radar cross section has been achieved.

Possible loads enumerated for the B-52 and new B-1 bomber aircraft are 12 externally and eight internally on the former type, and 24 carried internally by the supersonic B-1. The B-52's external ALCMs could be fitted with auxiliary fuel tanks for extended range, possibly to twice that of the standard version.

A nuclear warhead of the type carried by the follow-on SRAM Short Range Attack Missile will be used, and the guidance package is to be the same for both the Tomahawk cruise missile and ALCM. Guidance (overland) will be by the TERCOM (Terrain Contour Matching) technique which has been pioneered by the US Navy, and a terrain following system will be provided to enable penetration at a height of a few hundred feet.

The first powered ALCM flight took place on March 5, 1976. By early 1977 the advanced development programme had been completed and full-scale development had started. The original Intitial Operational Capability date of December 1981 was brought forward to December 1980 at that time, and Fiscal Year 1978 funding amounted to about $24m. Main contractor is the Boeing Aerospace Company.

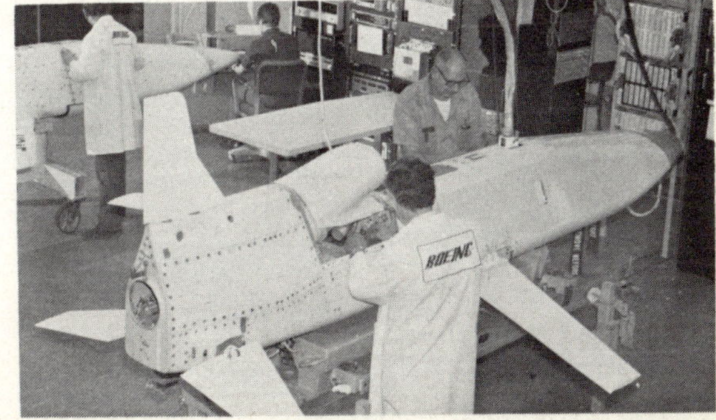

Exploded drawing (left) shows main components of ALCM, while picture on right taken at the Boeing plant indicates size of the weapon

AMRAAM

(USA)

The AMRAAM (Advanced Medium Range Air-to-Air Missile) programme is a joint USAF/USN project having the objective of providing a new generation of air-to-air missiles capable of providing new fighters such as the F-14, F-15, F-16 and F-18 with the ability to engage high performance enemy combat aircraft. The new missile will also supplant the AIM-7F Sparrow and AIM-54 Phoenix in some roles, and it is intended to be lighter and cheaper while incorporating a number of advances. This programme is also referred to as the BVR (Beyond Visual Range) missile requirement, parallel to which is a complementary project to develop WVR (Within Visual Range) missiles to replace the AIM-9L Sidewinder as the interim dog-fight missile with US air forces.

Northrop AMRAAM test vehicle firing

Five American companies are competing in the AMRAAM programme, General Dynamics, Hughes, Northrop, Ford Aerospace, and Raytheon. The first three have received government contracts, while the other two have provided their own funding. A three-year programme is expected to lead to the selection of two competing versions from which one will be selected for engineering development and eventual production in the mid-1980s.

A 'fire-and-forget' capability is a major requirement of AMRAAM, and this is likely to be achieved by the adoption of an active radar seeker head. Northrop have revealed some details of their AMRAAM project and these reveal the use of advanced aerodynamic techniques also. In April 1977, Northrop successfully completed ground-launched flight tests of an AMRAAM airframe at Fort Irwin and this design featured extremely high speed, manoeuvrability and low drag characteristics. Unlike other radar air-to-air missiles, no drag-inducing wings are employed; it is controlled by four surfaces on the tail and manoeuvres by means of body lift. In operation, the Northrop AMRAAM will use an inertial unit and micro-computer to project target co-ordinates obtained from the launch aircraft's radar. This will direct the missile to the target area where the active radar head will lock on to the target to guide the missile to impact.

In fiscal year 1977, Congress appropriated $5.0 million to start up the BVR missile programme, and the amount sought for the following year was about $42.5 million.

ANAB (AA-3)

Anab is the NATO code-name assigned to the Soviet air-to-air missile first seen carried by the Yak-28 (Firebar) in 1961. The US designation is AA-3. It has subsequently been seen on the Su-9 (Fishpot) interceptor, and is known to have been adopted as a standard weapon by the Soviet forces.

The existence of both radar and infra-red homing versions have been reported. Length of both versions is estimated at approximately 360 cm, diameter 28 cm, wing span 130 cm. Anab has a cylindrical body with a large cruciform wing assembly at the rear of the missile, and a set of four in-line fins ahead of the wings, and about one-quarter of the missile length from the nose end. Solid fuel propulsion is assumed, and the range has been quoted by the UK MoD as in excess of 16 km. The same source also estimated total production as certainly in the thousands.

This weapon has been in operational use for a considerable time and it is reasponable to assume that it has been the subject of periodic improvements and updates in the same way as some long-life Western counterparts have been modified. Some observers have claimed the existence of an Advanced Anab, but no details of any specific improvements have been obtained.

Type: Air-to-air
Configuration: Cylindrical body with tapered nose. Large cruciform wings at rear end and four small control surface forward
Length: 3.6 m Provisional
Diameter: 28 cm Provisional
Span (Max): 1.3 m Provisional
Propulsion: Solid
Range: 16 km Provisional
Guidance: Infra-red and radar versions
Warhead: High-explosive

APEX (AA-7)

Apex, AA-7, is one of three new-generation Soviet air-to-air missiles which became known in the West in 1976. It is thought to be a successor to Anab, to which there is some resemblance, though Apex is clearly of superior performance and somewhat larger. In particular, the provision of a supplementary set of control surfaces at the rear of the missile, in addition to the canard fins ahead of the wings, indicates high manoeuvrability. Both infra-red and radar homing versions of Apex have been reported and one of each kind comprises the standard armament of the interceptor version of the MiG-23. Solid propulsion is assumed.

Type: Air-to-air
Configuration: Cylindrical body with cruciform clipped delta planform wings. Small control surfaces at rear and forward of wings
Length: About 4.3 m Estimated
Diameter: About 24 cm Estimated
Span (Max): 1.05 m Estimated
Weight: 300-350 kg Estimated
Propulsion: Solid
Range: 15-30 km
Guidance: IR and radar versions
Warhead: HE up to 40 kg

APHID (AA-8)

Aphid, AA-8, is thought to be a small dogfight air-to-air missile, perhaps derived from the AA-2 Atoll but by no means inevitably so. Like the Apex, Aphid may be deployed on MiG-23 interceptor aircraft. Infra-red and radar homing models have been reported.

Type: Air-to-air, 'dogfight'
Configuration: Slender cylindrical body with cruciform long-chord wings at rear and control fins near nose
Length: 2.1 m Estimated
Diameter: 13 cm Estimated
Span (Max): 54 cm Estimated
Weight: 50 kg Estimated
Propulsion: Solid
Range: 8 km Estimated
Guidance: IR and radar versions
Warhead: HE about 7 kg

AS-6 (USSR)

Little is known about the AS-6 air-to-surface missile which is believed to arm the Soviet Backfire supersonic bomber. In 1976 a British MoD statement quoted a range of over 220 km, but gave no other information. It has been suggested that alternative conventional and nuclear warheads exist and that propulsion is by a liquid-fuelled rocket motor, but these details can only be regarded as provisional.

AS.15

The AS.15 is an Aerospatiale naval air-tosurface missile project, revealed in model form for the first time at the Le Bourget Exposition Navale in October 1976. Like the existing Sea Skua, the AS.15 would be used to arm light/medium naval helicopters for anti-ship operations. No other firm details are available.

AS.20

Though now obsolescent, the AS.20 has been delivered in substantial quantities and is thought to remain in the inventories of the French Air Force and Navy, the German and Italian Air Forces, and at least two other countries. Radio command guidance is used to direct and steer the AS.20 to the target, and the rear of the missile carries twin flares to aid tracking. The same system of control that is used for the AS.30 missile can also be used for the AS.20, and the latter weapon is proposed by the manufacturers for use as a training round for AS.30 operation. Launch of the AS.20 is possible from any aircraft capable of a speed of Mach 0.7 or over.

Type: Air-to-surface
Configuration: Cylindrical body with pointed nose. Swept cruciform wings
Length: 2.59 m
Diameter: 25 cm
Span (Max): 78 cm
Weight: 143 kg
Propulsion: Two-stage solid
Range: 7 km
Guidance: Radio command
Warhead: High-explosive, 30 kg
Main Contractor: Société Nationale Industrielle Aérospatiale

AS.30

(FRANCE)

This weapon is to a large extent a larger and higher-performance development of the AS.20 tactical air-to-surface missile. A more powerful motor permits launches from a lower speed, Mach 0.45, compared with the Mach 0.7 minimum launch speed of the AS.20. Optical target tracking with radio command link guidance is another common feature, but the TCA (Telecommande Automatique) system developed as an alternative for the AS.30 has since been applied to the AS.20 and AS.12 missiles. This technique relieves the operator of the task of tracking the missile, this being performed automatically by an infra-red seeker carried by the aircraft, which tracks flares on the missile. This data is related to the optical line-of-sight to the target to produce automatic guidance commands.

At the 1975 Paris Air Show another delivery mode was illustrated. Under the project name 'Ariel', Thomson-CSF and Aerospatiale are collaborating in the development of a laser-guided version of the AS.30 to arm Jaguar aricraft. Thomson-CSF, who have an agreement with Martin Marietta Aerospace, have designed a laser seeking homing head for the AS.30 (which is also suitable for use with any air-to-ground weapon of 100 mm or greater diameter). The French and American concerns have developed a complementary laser target designation pod called 'Atlis'.

In addition to the French Air Force and Navy, other AS.30 users include the forces of Israel, South Africa, West Germany, Switzerland and the UK. Licence production in India has also been reported.

Type: Air-to-surface
Configuration: Cylindrical body with pointed nose. Swept cruciform wings
Length: 3.79 m
Diameter: 33 cm
Span (Max): 1 m
Weight: 500 kg
Propulsion: Two-stage solid
Range: 10-12 km
Guidance: Radio command, with optional manual or semi-automatic control. Laser guidance under development.
Warhead: High-explosive. 230 kg
Main Contractor: Société National Industrielle Aérospatiale

AS.30 air-to-surface missile

ASALM

ASALM (Advanced Strategic Air-Launched Missile) is the title of a USAF technology programme directed towards the design and development of a successor to the current SRAM weapon carried by bomber aircraft. As currently configured the ASALM is about the same weight and size as SRAM and is equipped with the same nuclear warhead. Range is stated to be somewhere between that of SRAM and the ALCM (Ie. 160 km to 1,300 km), although this will be conditional upon the flight profile selected. Overall length is 4.26 m and the standard SRAM launcher rack can accommodate ASALM without modification. The power plant is an integral ramjet with sustainer fuel contained in the centre portion of the missile body, and the booster aft of this. The air inlet is beneath the radome nose. A dual-mode seeker is fitted which also has an air target capability. Inertial guidance is employed for the cruise phase.

The first fight tests are not expected to take place until 1978 or 1979. Raytheon and Rockwell International have both been awarded USAF contracts for guidance system studies, and McDonnell Douglas and Martin-Orlando are working on ASALM air-frame design and technology.

ASH (AA-5) (USSR)

Ash is the NATO code-name assigned to large missiles carried beneath the wings of Soviet Tupolev 'Fiddler' long-range interceptor aircraft. The American designation is AA-5. Four such missiles, which are assumed to be air-to-air weapons, can be carried by each aircraft. Both radar guided and infra-red homing versions are in use, two radar homing missiles generally being carried on the wing outer pylons and two infra-red homing missiles on the inner pylons of the Fiddler interceptor. Estimated dimensions are : length 330 cm, wing span 130 cm, diameter 30 cm. Cruciform wing and tail surfaces, in-line, are fitted, the former having a sharply swept delta planform. The tail fins have little sweep on either leading or trailing edges and are mounted close behind the wing trailing edges.

The size of the missiles, parent aircraft, and the large nose radome of the latter, suggests that they are intended for long-range interception. A British MoD report of April 1976 gave a figure of about 30 km as the range. It was also stated that some thousands had been produced.

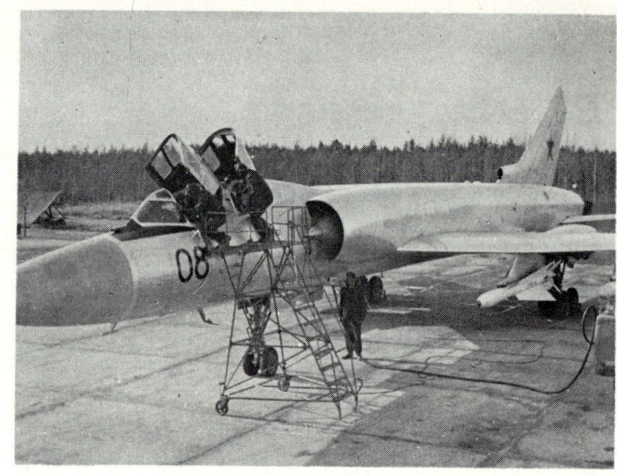

Russian AA-5 Ash air-to-air missiles

Under the designation ASM-1, the first Japanese tactical air-to-surface anti-ship missile is being developed for use with the Mitsubishi F1 close-support aircraft. No official details have been obtained but provisional characteristics are range approximately 45 km, speed about Mach 1, and it has been reported that a 140kg warhead will be carried. It has been officially stated that the ASM-1 will also be capable of ship or ground launching by the addition of a rocket booster motor. An inertial system will be used for mid-course guidance and an active radar seeker for terminal homing. In November 1973 Mitsubishi Heavy Industries Ltd was selected as prime contractor for the development of the ASM-1. Development is expected to be completed during 1979 with production possibly resulting in the following year. First ground firings may take place in 1977 but air launches are not expected before 1978.

ASPIDE

(ITALY)

Aspide is a high performance multi-role missile for operation at all altitudes against a wide number of target types, and is optimised for air-to-air and surface-to-air missions. It is based on the use of a semi-active radar, with monopulse receiver, guidance system which will ensure all-weather and all-aspect operation. It was first shown in public at the 1974 Hanover Air Show when it was seen to have a configuration closely resembling the Sparrow missile it is to replace. Principal characteristics are length, 3.7 m, body diameter 20.3 cm, wing span 100 cm (air-to-air) or 80 cm (surface-to-air), fin span 80 cm (air-to-air) or 64 cm (surface-to-air), launch weight 220 kg, burn-out weight 163 kg. Propulsion is by means of a single-stage solid-propellant motor which will give speeds of more than Mach 2.5 in the surface-to-air role.

Present plans are to employ the Aspide on interceptor aircraft, in the Albatros naval point defence weapon system; and in the 'Spada' ground-based low altitude air defence systems. In the first two systems, Aspide will replace the Sparrow III (AIM-7E) missile currently employed.

Further details of land and naval applications of Aspide will be found in the relevant sections of this book. Production deliveries were planned to being in 1977.

Air-launched version of Italian Aspide multi-role missile

Atoll is the NATO code-name assigned to a Soviet air-to-air missile, believed to bear the USSR designation SB06 and/or K13A, and for which the US designation is AA-2.

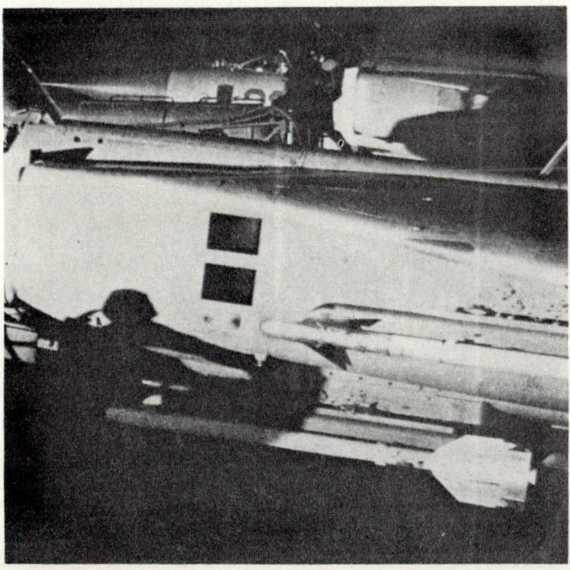

The missile closely resembles the American AIM-9B, infra-red homing Sidewinder and is of similar dimensions and (estimated) weight. Atoll is widely deployed with MiG-21 (Fishbed) interceptors of the Soviet home forces and on export versions of this aircraft. Known foreign users include Egypt and India. The latter country, which has over 50 MiG-21s, has facilities for Atoll production under licence. In addition to the Warsaw Pact countries, there is an impressive list of foreign users which includes: Afghanistan, Algeria, China, Cuba, Egypt, Finland, India, Iraq, North Korea, Syria, and Vietnam.

An advanced version has been reported, and it is probable that during a long operational life it has been subject to numerous improvements in the same way as the broadly equivalent American Sidewinder. In addition to the Indian licence-built version, there is almost certainly an indigenous Chinese model.

Type: Air-to-air
Configuration: Slim cylindrical body with large tail surfaces and four control vanes near nose
Length: 2.8 m
Diameter: 12.7 cm
Span (Max): 53 cm
Propulsion: Solid
Guidance: Infra-red
Warhead: High-explosive

AA-2 Atoll air-to-air missiles loaded on a Soviet interceptor

BRAZO (USA)

The designation Brazo (Spanish for 'arm') refers to a US Navy project for the development of an air-to-air defence missile that will home on to the radars of attacking aircraft such as the Soviet Foxbat high-speed, high-altitude fighter. Since the programme was made public in mid-1972, the USAF has joined in and is responsible for test and evaluation. It has named the project Pave Arm.

The US Naval Electronics Centre is designing a passive broad-band anti-radiation homing head, several versions are known to have been produced for the missile which is based on the Sparrow. Hughes Aircraft is responsible for the integration of the new heads with a quantity of Sparrow missiles which are being used for test and demonstration flights. A BQM-34 drone was intercepted in the first air launch from an F-4D in April 1974. The first free flight of an unarmed Brazo test vehicle was made from an F-4 fighter aircraft on April 16, 1974 at Holloman AFB. The target was a BQM-34 drone. A second series of test firings began in October 1974 and ended in January 1975. In a launch from a USAF F-4D Phantom in October 1974, Brazo achieved its first long-range kill when it destroyed a BQM-34A target in a head-on, look-down interception. In following tests Brazo scored three hits from three launches to complete the programme's initial demonstration phase. The first of the three was a look-down, tail attack, and the second and third firings were long-range, look-down nose interceptions.

BULLPUP (AGM-12)

The Bullpup AGM-12 series embraces a range of air launched missiles for attacking tactical surface targets on land or sea. Guidance is by radio command link from the launching aircraft, with optical tracking to target, aided by flares attached to the missile. Bullpup model designations run from AGM-12A to AGM-12E, and within the group quite considerable variations of payload and dimensions exist. Early versions of Bullpup employed a US Navy-developed solid fuel rocket motor, but all current models use pre-packed liquid fuel motors (LR62, LR58, Thiokol Reaction Motors) which permit long storage time and rapid missile preparation. The several variants of Bullpup carry either 250lb (113.4kg) or 1000lb (453.6kg) conventional high explosive warheads, or a nuclear warhead.

Bullpup became operational in 1959 and is now in use by the US Navy, Air Force, and Marine Corps, and numerous NATO services. Foreign users include the Royal Navy which selected the AGM-12B as standard strike armament for the Buccaneer aircraft. This version of Bullpup is manufactured under licence in Europe for the armed forces of Denmark, Norway, Turkey, and the UK by a consortium headed by Kongsberg Vaapenfabrikk. American production has ceased.

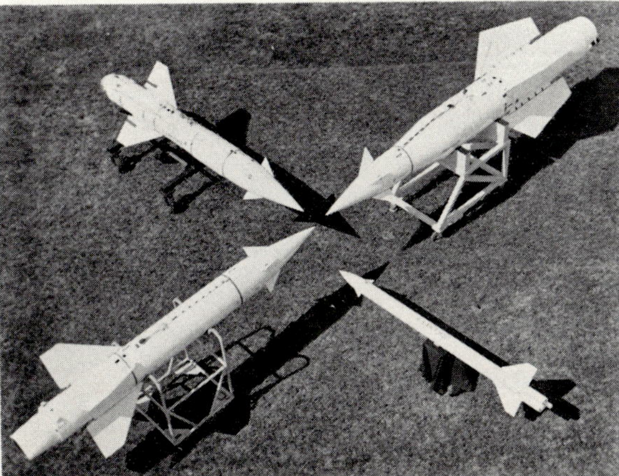

	AGM-12B	AGM-12C
Type:	Bullpup-A	Bullpup-B
Length:	3.2 m	4.16 m
Diameter:	30.5 cm	46 cm
Span (Max):	1 m	1.22 m
Weight:	258 kg	810 kg
Propulsion:	Pre-packed liquid	
Range:	11 km	17 km
Guidance:	Radio command	
Warhead:	HE 113 kg	HE 454 kg

Main Contractor: Martin Marietta Aerospace.

(Bullpup AGM-12D has same airframe as AGM-12C but carries a nuclear warhead.)

Bullpup family. Clockwise from top right: AGM-12C, ATM-12A (training round), AGM-12B, and AGM-12D (nuclear)

EXOCET (AM.39) (FRANCE)

A special version of the successful Exocet anti-ship missile was developed under the designation AM.39 and this model is to arm Super-Frelon helicopters of the French Naval Air Arm. The basic missile is similar to the MM.38 version described in the Naval Missile Systems section of this book, but is somewhat shorter as the table there shows. The improved motor used in the MM.40 Exocet is incorporated in the AM.39. The Super-Frelon is equipped with an Omera ORB-31D radar for target detection and missile direction. A high-altitude, fixed-wing aircraft variant of the AM.39 Exocet has been studied for the Atlantic and Super-Etendard aircraft. Pakistan and South Africa are reported to have ordered the air-launched Exocet.

Launch of AM39 Exocet anti-ship missile

FALCON (AIM-4, AIM-26, and AIM-47)

The Hughes Falcon family of air-to-air missiles is a long-lived and prolific one, extensive enough to have undergone sufficient stages of evolution to warrant the three basic US designations given in the heading to this entry. These, in turn, have progressed through a number of variants denoted by suffix letters. The main features are outlined in the following paragraphs.

Falcon was developed jointly by Hughes and the USAF, work starting on the project in 1947. An experimental model was produced two years later. First production models appeared in 1954. The Falcon was introduced as the GAR-1 and about 4,000 were built before an improved version appeared as the GAR-1D (now AIM-4A). Over 12,000 of this model were produced. It was equipped with a semi-active radar homing head operating in conjunction with a target illuminating radar carried by the launching aircraft. Powered by solid fuel rocket motor. High explosive warhead. Principal characteristics are: length 198 cm, diameter 16.25 cm, wing span 50.8 cm, weight about 54.4 kg. In 1956 the AIM-4C was introduced, being essentially the same as the AIM-4A but with an infra-red homing head.

Chronologically, development of the AIM-4D took place after that of other members of the Falcon series, such as AIM-4E, F, and G and its former designation was GAR-2B. As air defence missiles, earlier Falcon versions were designed for the interception of bomber aircraft. AIM-4D development was undertaken to improve performance sufficiently for the engagement of enemy fighter aircraft. Principal differences from the AIM-4C are adoption of the improved infra-red homing head of the AIM-4G Super Falcon, thereby providing an all-aspect attack capability. Many A and C models were converted to this standard.

The AIM-4E was originally developed as the GAR-3 and was introduced in 1958. It was succeeded two years later, after 300 units had been produced, by the AIM-4F (formerly GAR-3A).

Falcon family. Foremost is the AIM-4D, in the background (l to r) are: AIM-½G, AIM-4E, AIM-4C, and AIM-26A

Guidance is by semi-active radar homing as in the AIM-4A. The AIM-4E was powered by a longer burning solid fuel rocket motor to provide longer range, a higher launching speed, and a higher combat ceiling than earlier models. A more powerful high explosive warhead was also fitted. Only about 300 models of this version were produced before being succeeded by the later variants, the AIM-4F and AIM-4G

The AIM-4G Super Falcon is the infra-red seeking counterpart of the AIM-4F missile. It is equipped with an I/R detector system which enables it to lock-on to smaller targets at greater ranges. The same seeker is used in the AIM-4D Falcon. Compared with the AIM-4F Super Falcon, the AIM-4G is shorter (105.7cm) and weighs slightly less (65.7 kg). A high-explosive warhead is fitted. Unofficial weight is 18 kg, and speed Mach 3. Developed under the original designation GAR-4A, in parallel with the GAR-3A (AIM-4F), it was introduced in 1959/60.

An experimental series of missiles XAIM-26A (formerly XGAR-11) was introduced in the Spring of 1960. These models, while bearing a family resemblance to earlier Falcons, presented a very much bulkier appearance and were in fact much heavier — about 90 kg. Length was 213.4 cm and maximum diameter 27.9 cm. Wing span remained about 60 cm. Radar guidance was employed and a nuclear warhead was carried. Extra weight was accounted for by a larger motor and heavier warhead, the former being required for increased range. An active proximity fuze, suitable for 'near-miss' kill capability was also fitted and this is believed to have been based on radar. Four fuze aerials are located almost flush with the missile body, ahead of the wings. The warhead was essentially the same as that of the nuclear Genie missile (which see). The XAIM-26A was followed by the AIM-26B (formerly GAR-11A), a conventional explosive warhead version of the AIM-26A.

AIM-47A was originally developed under the designation GAR-9 for the USAF as part of the YF-12A Mach 3 defence interceptor programme. This is the largest of the Falcon series and is about twice the size and weight of the AIM-4 series. Estimated dimensions are: length 320 cm, diameter 33.5 cm, wing span 83.8 cm, weight 363 kg. A solid-propellant motor is used and the AIM-47A has been credited with a range of 100 nm and a speed of Mach 6. Guidance is semi-active radar homing with the Hughes AN/ASG-18 fire control system radar acting as target illuminator. It is reported that either nuclear or conventional payloads can be carried.

Falcon missiles have been extensively supplied to overseas air forces, and in certain cases licence production is undertaken. Export designations are HM-55 for the AIM-26B which is built in Sweden as the RB27, and HM-58 is similar to the AIM-4C/D being built in Sweden as the RB28.

FIRESTREAK

Development of Firestreak was started in 1951 and the first fully guided round was fired in 1954. In the following year, the first trial firing against an airborne target was made at the Aberporth missile range in Wales. Entry into service with the RAF and Fleet Air Arm of the Royal Navy occurred in 1958 and Lightning, Javelin and Sea Vixen aircraft were among the types to be equipped.

Production was completed in 1969 but Firestreak remains in service with RAF, Kuwaiti, and Saudi Arabian Lightnings.

Type: Air-to-air
Configuration: Cylindrical body with cruciform wings aft of mid-length point and four small control surfaces near tail
Length: 3.19 m
Diameter: 22 cm
Span (Max): 75 cm
Weight: 136 kg
Propulsion: Solid
Range: 1.2-8 km
Guidance: Infra-red
Warhead: High explosive, 23 kg
Main Contractor: Hawker Siddeley Dynamics Ltd

Firestreak on an RAF Lightning interceptor

GBU-15(V) (USA)

Later developments of the electro-optical (EO) and laser homing 'smart bomb' systems have been gathered together within a USAF programme designated the GBU-15(V), Modular Glide Weapon System. GBU-15(V), is a variable designator used for a family of weapons modules. These can be configured in various combinations to provide for various attack and target conditions. The base module set comprises those items needed to convert standard munitions, such as the MK 84 bomb, to EO smart bombs of the sort described in the later entry on HOBOS, and which were extensively used in SE Asia.

The full list of major components comprises:

(1) Three warhead options: Mk 84 'iron' bomb; SUU-54 dispenser; or HSM (Hard Structure Munition).
(2) Three wing and tail surface packages: small wing module; cruciform wing module; and planar wing module.
(3) Four guidance modules: EO; Laser; Imaging Infra-Red; and DME.
(4) Two adaptors to join the warhead to the guidance package.
(5) A control module.
(6) A data link module.

The four guidance modules are interchangeable so that a weapon can be provided for specific conditions — day, night, high- or low-visibility attack, or all-weather attack. The EO seeker is essentially the same as was used in SE Asia. This is a TV seeker and is used for attacks where the pilot can see the target prior to launching the weapon. This is called a direct attack. By adding the data link module, an indirect attack is possible.

The laser and Imaging Infra-Red seekers are those being developed for the Maverick programme (which see), the former for use when conditions do not allow use of the direct attack TV weapon and the latter for night operations. Both warhead modules weigh 2,000 lb (907 kg), the Mk 84 being intended for targets such as railways, buildings and bridges, while the SUU-54 dispenser is intended for use against area targets such as surface-to-air missile sites. The SUU-54 disperses 1,800 grapefruit-size bomblets.

The three types of wing sets are interchangeable also to cater for both low-altitude and high-altitude attacks. The cruciform wing is being developed to improve on the small wing configuration used in Vietnam, and this design gives the weapon increased manoeuvrability at low altitudes. When used with the HSM warhead, the cruciform wing set can have booster rocket motors attached to the tips of the rear fins wings to increase impact velocity. The planar wing is being developed to provide a high-altitude weapon that will have an appreciable range when launched at altitude. When used with the DME guidance unit, a data link module and the EO seeker, this weapon will permit attacks from stand-off ranges.

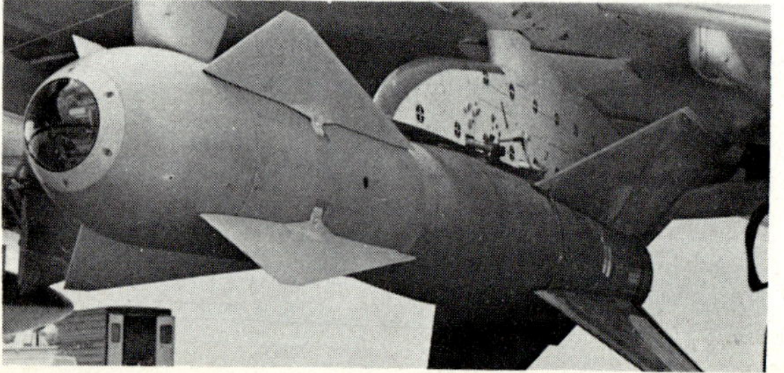

USAF GBU-15 cruciform wing Modular Glide Weapon

GENIE (AIR-2A)

Development of Genie was started in 1954 and the first successful tests from aircraft launches were conducted in 1956, probably making Genie the first nuclear air-to-air weapon when it entered USAF service in January 1957. Prior to this stage the Genie programme had been known by various code-names, among them Ding-Dong and High Card. A subsequent programme for the development of a guided version, AIR-2B or Super Genie was later dropped in 1963. Genie production ended in 1962 when a total of several thousand units had been delivered. One estimate puts a figure of 10,000 on total output. A proposal to adapt Genie for air-to-surface missions, devised by the USAF at Hill AFB in 1966, failed to materialise.

Type: Air-to-air
Configuration: Circular cross-section body with fore part of slightly bulbous shape. Four stabilising fins at rear end
Length: 2.74 m
Diameter: 43 cm
Span (Max): 60.9 cm
Weight: 372 kg
Propulsion: Solid 16,330 kg thrust
Range: 9.6 km
Guidance: Unguided. Gyro stabilised
Warhead: Nuclear
Main Contractor: McDonnell Douglas Astronautics Company

*Genie nuclear air-to-air missiles awaiting loading. (*USAF*)*

HARM (AGM-88A) (USA)

The HARM programme has as its objective the development of a new anti-radiation missile for use in defence suppression and similar operations, and which offers performance improvements over the existing Shrike and Standard ARM. Higher speed, faster reaction, and a more destructive warhead are the likely priority improvement areas. Other desired characteristics which have been specified are: relatively low cost, integrated avionics and Electronic Warfare systems, integrated circuit construction, latest ECCM, weight of about 350 kg, an improved propulsion, system, and improved frequency coverage.

First contracts in support of HARM research and development were awarded during 1972. Among them were the following: Texas Instruments, guidance and avionics; Hughes Aircraft, systems analysis and guidance support; Itek, AN/ALR-45 radar warning receiver modifications; Lockheed, system studies; Dalmo-Victor, modification of DSA-20N signal analyser; Stanford Research Institute, analysis. In late 1974 Texas Instruments was selected as weapon system integration contractor for HARM. Present plans call for completion of R and D in 1980.

Type: Air-to-surface anti-radiation missile
Configuration: Not yet defined
Length: 4.17 m
Diameter: 24 cm
Span (Max): 1.13 m
Weight: 354 kg
Propulsion: Probably solid
Range: 16 km
Guidance: Passive, broad-band RF radiation homing
Warhead: High-explosive
Main Contractor: US Naval Weapons Centre

*HARM, High-speed Anti-Radiation Missile (*left*) and Shrike side-by-side on A-7E Corsair II aircraft. (*USN*)*

(USA) HELLFIRE

Hellfire is a helicopter-launched fire-and-forget modular missile system to meet the US Army requirement for an air-to-ground terminal homing missile to engage single hard point targets such as tanks. The modular missile is designed to accept a succession of homing seeker modules, as they evolve. As presented in early 1974 the Hellfire system programme consisted of advanced development of a laser Hellfire and, as a supplementary effort, the advanced development and integration of a radar frequency/infra-red (RF/IR) air defence suppression seeker.

The current plan envisages a Hellfire missile able to accept any terminal homing seeker (laser, TV, IR, RF, or dual-mode RF/IR). The laser seeker is being developed by the USAF as lead service and incorporates Air Force and Army specifications to satisfy both Maverick and Hellfire requirements. Rockwell International was awarded a $67m Hellfire development contract by the US Army in October 1976. Fiscal Year 1978 funding for laser Hellfire was $50.5m.

Hellfire missiles on US Army Cobra helicopter

HOBOS

(USA)

There are two types of smart bombs currently in service with the USAF — the laser-guided bomb first introduced in Southeast Asia in May 1968, and electro-optically (EO) or HOBO bombs which entered service in that theatre in February 1969. The former type of munition is used in 500, 2,000 and 3,000lb categories and the EO bomb is produced in 2,000 and 3,000lb sizes.

The laser-guided type consists of a conventional bomb to which are added stabiliser fins, a computerised directional package in place of the normal fuze, and a guidance module containing a laser seeker. (Further details will be found in the Paveway entry, see page 223) The EO bomb has strakes running the length of the weapon and control surfaces at the rear, with a TV guidance module at the front. A monitor in the aircraft is used by the crew for initial target acquisition, and after lock-on by the bomb's camera has been confirmed the weapon can be released, after which the guidance module steers the bomb to the target automatically. The laser-guided versions are reported to cost $3,100 each, and the EO versions $13,000 each. In both cases CEP errors are claimed to have been reduced to a few feet.

In the latter half of 1972 the USAF, with industry, instituted studies of possible ways of further enhancing the smart bomb concept by the provision of blind attack facilities and extending the effective range to permit 'stand-off' operations. This became the GBU-15(V) Modular Glide Weapon System programme, described above.

Type: Guided bomb
Configuration: Mk 84 (907 kg) or M 118 (1,360 kg) bomb modified by addition of an electro-optical guidance package and four control surfaces at rear

	Mk 84	**M 118**
Length:	3.78 m	3.70 m
Diameter:	46 cm	61 cm
Span (Max):	112 cm	132 cm
Weight:	1016 kg	1544 kg

Propulsion: None
Range: Dependent on release height and speed
Guidance: Electro-Optical contrast tracker or I/R seeker. Pneumatic tail control actuation. Roll stabilised. Proportional navigation.
Warhead: HE. Mk 84: (907 kg). M 118: (1360 kg)
Main Contractor: Missile Systems Division, Rockwell International

HOBO, electro-optical homing bomb

HOUND DOG (AGM-28B)

The AGM-28B, Hound Dog is a long-range, stand-off air-to-ground strategic missile and is carried in pairs beneath the wings of USAF Strategic Air Command B-52 aircraft. The prime navigational mode of the Hound Dog missile from launch to target is an inertial navigation system. Prior to launch, missile INS remain operative and are continuously updated by the navigation system of the parent B-52. Missile targeting and mission profile can be varied by the B-52 crew in flight.

Development was initiated by USAF Strategic Air Command in 1957, and the first launch took place in 1959. Entry into service under the original designation GAM-77A was in 1961. The AGM-28B (formerly GAM-77B) designation was allocated to an improved version. Production ended in 1963. In 1974 the then US Secretary of Defence announced that it was planned to phase out Hound Dog by 1976, but it is thought that implementation of this may have been deferred.

Type: Strategic, air-to-surface
Configuration: Long slender cylindrical body with gently tapered nose. Underslung podded engine aft. Two delta shaped fore-planes, and aft-moounted delta wings. Vertical fin and rudder
Length: 12.95 m
Diameter: 71 cm
Span (Max): 3.66 m
Weight: 4500 kg
Propulsion: Turbojet J52-P-3 engine
Range: 965 km
Guidance: Inertial, updated to time of launch by parent aircraft navigation system and astro-nav system in launch pylon.
Warhead: Nuclear. Provisional yield 4 megatons
Main Contractor: North American Aviation, Inc

Two Hound Dog missiles can be carried by B-52 bombers. (USAF)

KANGAROO (AS-3) (USSR)

Kangaroo is the NATO code-name assigned to the largest Soviet air-to-surface missile so far revealed. The US code number is AS-3. First shown in public in 1961, it is associated with the Tu-95 Bear bomber aircraft and is generally assumed to have retained operational status though its operational importance has probably declined significantly as newer weapons have been deployed. Kangaroo has the configuration of a conventional swept-wing fighter aircraft and similar dimensions, approximately 15 m in length and with a span of 9 m. The propulsion used has not been confirmed but is probably a turbojet engine. There is no reliable information available as to the ratio in which the relatively large payload potential of Kangaroo is allocated between range and warhead, as is borne out by the wide variation in range estimates published by different specialist sources. These extend from 185 km to 650 km, and it is likely that all these ranges are possible but without access to trustworthy data of the type and weight of warhead, guidance method(s) and performance, the true range figure must remain conjecture.

Similar considerations apply to Kangaroo's military function, but the most probable is that the weapon was designed as a medium/long-range stand-off nuclear weapon for use against 'area' targets such as cities or industrial areas, rather than point targets.

Type: Strategic, air-to-surface
Configuration: Similar to conventional swept-wing fighter aircraft, with swept tail surfaces
Length: 15 m Provisional
Span (Max): 9 m Provisional
Propulsion: Probably turbojet engine
Range: Up to 650 km Provisional
Guidance: Probably auto-pilot
Warhead: Probably nuclear

KELT (AS-5)

Kelt is the NATO code-name of the AS-5 Soviet air-to-surface weapon, first seen on a Tu-16 Badger bomber aircraft. It has superficial but pronounced similarities to the Kennel, which also is used with the Badger, and with the Styx shipborne surface-to-surface missile. The fuselage centre-body and wings of Kelt appear to be the same or very similar to those of Kennel, while the nose section (and presumably the guidance system it carries) of Styx is employed. A rocket propulsion motor is used and the tail control surface arrangements of Kelt differ from both those of both Styx and Kennel. Western estimates of range vary between 160 km and more than 320 km, but a British MoD report issued in April 1976 quoted the former figure as maximum range for the AS-5 (Kelt). Total production is over 1 000 missiles.

Type: Air-to-surface. Tactical
Configuration: Aircraft type configuration. Circular cross-section fuselage with radome nose. Swept wings and tail surfaces
Length: 9.4 m
Diameter: 1 m
Span (Max): 4.6 m
Propulsion: Unidentified rocket motor
Range: Probably over 160 km
Guidance: Probably command plus active radar homing
Warhead: High-explosive or nuclear could be carried

Kelt, AS-5, missiles are carried beneath wing of this Soviet bomber

KENNEL (AS-1)

Thought to be the earliest of the Soviet stand-off weapons, the AS-1 Kennel missile is carried by the Tu-16 Badger bomber, one beneath each wing. In general configuration very much like a scaled-down MiG-15 fighter aircraft, it has a prominent radome superimposed on the upper rim of theair intake. Other external evidence of electronics carried by Kennel suggests that beam-riding radar or radio command link guidance is used for a large portion of the trajectory, with either passive or active radar homing for the terminal phase. The parent Badger aircraft are provided with at least two radars on the underside of the fuselage that could serve such a mode of operation. In addition to the Soviet forces, the Badger/Kennel combination has been supplied to the Indonesian and Egyptian Air Forces. The latter is reported to have used a Kennel in action on at least one occasion during the Middle East conflict of Autumn 1973. Apparently identical with the Samlet coast defence missile.

Type: Air-to-surface. Tactical, anti-shipping weapon
Configuration: Aircraft type configuration with fuselage of circular cross-section. Nose air intake with radome superimposed on upper rim. Swept wings. Tail plane mounted on sides of vertical fin and rudder
Length: 8.5 m Estimated
Span (Max): 4.9 m Estimated
Propulsion: Turbojet
Range: 90 km Estimated
Guidance: Probably command with either semi-active or active radar homing
Warhead: High-explosive

KERRY (AS-7) (USSR)

Very litle is known of the AS-7 Kerry air-to-surface missile, which is one of the new generation of Russian weapons reported in the mid-1970s. Its role is thought to be tactical and radio command guidance has been suggested. Carrier aircraft that have been named include the Su-17, Su-20, and MiG-23 support aircraft.

KIPPER (AS-2)

Similar in general configuration to the American Hound Dog stand-off missile, the Soviet AS-2 Kipper probably has a similar strategic bombing role although Western observers generally assign an anti-shipping role to this missile. It is normally deployed with Tu-16 twin-jet bombers of the Soviet Naval Air Arm, which supports the anti-ship mission. Kipper was first seen at the 1961 Soviet Aviation Day display at Tushino Airport, Moscow.

Type: Air-to-surface
Configuration: Similar to small fighter aircraft with swept wings and tail surfaces. Underslung pod-mounted engine at rear
Length: 9.5 m Provisional
Span (Max): 4.6 m Provisional
Propulsion: Turbojet engine
Range: 210 km Provisional
Guidance: Probably auto-pilot, possibly with radar terminal homing
Warhead: Possibly nuclear

KITCHEN (AS-4)

Kitchen (US code, AS-4) appears to be one of the most technically advanced Soviet air-to-surface missiles yet revealed. It has a fuselage just over 11 m long to which are attached a pair of delta planform wings and a cruciform tail assembly. Propulsion is by rocket motor, liquid-fuel burning according to American sources, and inertial guidance is assumed and the aerodynamic features of Kitchen suggest a high cruising speed. A UK MoD statement in early 1976 gave a maximum range of 298 km. The only Soviet aircraft type known to be equipped to carry this weapon is the Tu-22 Blinder, beneath which it was first seen in 1967. The missile is carried in a partly enclosed position on the fuselage undersurface.

Type: Air-to-surface
Configuration: Aircraft type fuselage with short-span delta wings and cruciform tail assembly
Length: 11.3 m Provisional
Propulsion: Liquid fuelled (Provisional) rocket motor
Range: Estimated maximum 300 km
Guidance: Auto-pilot/inertial
Warhead: Possibly nuclear

KORMORAN (AS.34) (W. GERMANY)

Kormoran (AS.34) is an advanced air-to-surface missile, designed principally for strikes against surface shipping. The guidance system employed provides a useful stand-off capability for the launch aircraft. Propulsion is by solid-fuel motors, and a hybrid guidance system is employed. The Kormoran is roll stabilised and is steered by aerodynamic control surfaces at the rear. The missile is constructed in three main sections: the nose, containing the homing head; warhead; and motor, which also includes the rear structure and the main navigation package that is carried ahead of the sustainer motor. The 160kg warhead is designed to penetrate hulls without being damaged, and detonation is initiated by an impact fuze with a pyrotechnic delay set to explode the warhead approximately in the centre of the target.

The navigation package consists of a twin-gyro platform, an inertial computer, associated electronics, and a time switch. There is a separate course and position computer in the tail section and also a radar altimeter. The radar homing head fitted is based on the RE 576 developed by Thomson-CSF. This is stated to be capable of either active or passive homing.

So far the West German Navy is the only named user for the Kormoran, but interest from the Italian Air Force has been reported, and use with the Tornado MRCA is also expected. Production has started and deliveries had started by the end of 1976. An initial production of 350 has been reported.

Type: Air-to-surface, anti-ship weapon
Configuration: Cylindrical body with conical nose. Cruciform cropped delta wings and small cruciform tail surfaces
Length: 4.4 m
Diameter: 34 cm
Span (Max): 1 m
Weight: 600 kg
Propulsion: Two-stage solid
Range: 37 km
Guidance: Inertial cruise, active radar homing. RE 576 homing head said to be capable of passive mode operation also
Warhead: High explosive, 160 kg
Main Contractor: Messerschmitt-Bölkow-Blohm GmbH

Kormoran air-launched anti-ship missile

MAGIC R.550

The Matra R.550 Magic air-to-air missile is designed for 'close-combat' operations (from less than 500 m to more than 6 km), with consequent emphasis upon the ability to withstand high load factors imposed by the severe manoeuvre demands required. Propulsion is by a solid rocket motor, and operational range is reported to cover from 200 m or less, to as much as 10 km. Twist and steer control of the missile is effected by the canard arrangement of fins at the forward end of the weapon. A SAT infra-red homing head provides guidance, and the special launch mount houses a liquid nitrogen container to provide cooling for the IR sensor.

The R.550 was first revealed at the 1971 Paris Air Show and was at the early development stage at that time. It is planned to replace Sidewinders with this missile on French Air Force and Navy interceptors. In September 1974 evaluation by the French Air Force was started and deliveries of the first series produced missiles also began at that time, continuing into early 1975. By the end of 1976, 1,000 missiles had been delivered. Both the French Air Force and Navy have the R.550 in their inventories. Orders for several thousand missiles have been placed by foreign countries, including Oman, Saudia Arabia, South Africa, and possibly Spain, Kuwait, and Greece. Two as yet unnamed South American states are also thought to have bought the R.550.

Type: Air-to-air dogfight weapon
Configuration: Cylindrical body with three sets of cruciform fins and rounded nose. Large rectangular fins at rear, double canard arrangement at fore end
Length: 2.93 m
Diameter: 16 cm
Span (Max): 65 cm
Weight: 90 kg
Propulsion: Solid
Range: 500 metres to over 6 km
Guidance: Infra-red
Warhead: High explosive
Main Contractor: Engins Matra

Matra R.550 Magic 'dog-fight' missile

MARTE

(ITALY)

The purpose of the Marte system is the destruction or disabling of naval craft in 'all weather' conditions by means of Sea Killer missiles launched from helicopters in a stand-off position. The helicopter-borne missile used in conjunction with the Marte Project is the Sea Killer MK.2 (20km range with 70kg warhead for the Marte version). Easily operated and light equipment only need be carried on board the helicopter since the radar performs navigation search and track of the target as well as guidance of the missile in azimuth (the missiles are autonomous in altitude by means of SISTEL-designed radar altimeter).

Installation of the Marte Weapon System has been studied on Agusta helicopters ranging from a maximum TO weight of 3,000 kg up to 10,000 kg or more depending on the number of missiles carried and on the operational requirements which may call for simultaneous ASW and Surface Strike capability or not. A system for the SH-3D helicopter is the subject of a pre-production contract from the Italian Navy.

Marte system test installation

MARTEL (AS.37 & AJ.168)

Martel is an air-to-surface tactical missile with two alternative terminal guidance systems capable of offering a considerable stand-off capability. Both versions are designed to operate in an ECM environment, to which a high resistance is claimed. The two forms of terminal guidance are: passive homing onto electromagnetic radiation in the AS.37, anti-radar version; and visual guidance to a selected target by means of a nose-mounted TV camera and a data link over which both video and command signals are passed between aircraft and missile, in the AJ.168 variant. The Martel system is the product of a joint Anglo/French development programme with prime responsibility for the AS.37 resting with Engins Matra, and for the AJ.168 with Hawker Siddeley Dynamics. Production contracts were placed by the two governments in December 1968. Martel will be used on Mirage IIIE, Jaguar, Atlantic, Harrier, and Buccaneer aircraft.

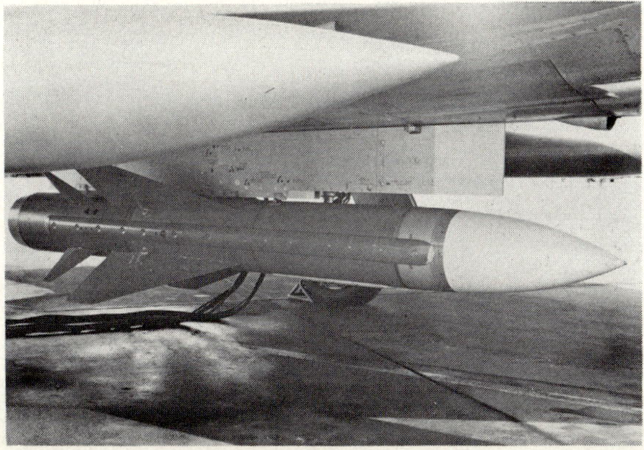

Subsequently, HSD based a number of developments on Martel. One was intended to meet a Royal Navy requirement for an Underwater-to-Surface Guided Weapon (USGW), but this was dropped in favour of buying the Harpoon from America in September 1975. A year later the Active Radar Sea-Skimming Martel (ARSSM) project was announced and limited government funding for this was given. More recently an air-breathing derivative powered by a Microturbo TI.60.1 turbojet was revealed and the Marconi active radar homing head for the ARSSM will probably be used. Possibly some or all of this work may be embodied in a future European anti-ship missile.

Type: Air-to-surface
Configuration: Cylindrical body with swept cruciform wings and four tail surfaces. AS.37 anti-radar version has pointed nose. AJ.168 TV-guided version has blunt nose
Length: AS.37, 4.2 m; AJ.168, 3.9 m
Diameter: 40 cm
Span (Max): 1.2 m
Propulsion: Solid
Range: 60 km Provisional
Guidance: Autopilot cruise. TV plus radio command for AJ.168; passive radar homing for AS.37
Warhead: High-explosive
Main Contractors: Engins Matra and Hawker Siddeley Dynamics Ltd

The anti-radar, AS-37, version of Martel

AIR FORCE
11369

Maverick is a relatively small, TV-guided tactical missile designed for use against small concentrated targets such as armoured vehicles, concrete field fortifications, revetments, gun positions, parked aircraft and communications vans. The missile is guided by a miniature TV homing system in the nose of the missile. The pilot picks up the target on a TV monitor in the cockpit, locks the missile's electro-optical tracker onto the target (by either slewing the missile's nose-mounted TV camera, or changing the aircraft attitude) and launches the missile, which is automatically guided to the target by the TV centroid tracker. The launch aircraft is free to take necessary evasive action once the missile is launched.

The basic missile is the subject of five enhancement programmes designed to either improve existing capabilities or to provide extra facilities. The salient features of the five models are briefly described in the following paragraphs.

Scene Magnification Maverick (AGM-65B):

This model is already in production. Its main advantage is that by use of new optics a stronger gimbal mount and slight changes in the electronics, the launch aircraft is able to depart from the defended area at greater slant ranges than before.

Laser Maverick (AGM-65C):

Laser Maverick will provide close air support against laser-designated targets in support of ground forces. It also provides for delivery of Maverick against low contrast and unbounded targets (i.e. those lacking well-defined visual contrast features) unsuitable for TV Maverick, by using either an airborne or a ground-based laser designator to define the aiming point. First flights of Laser Maverick took place in early 1977.

Infra-Red Maverick (AGM-65D):

This version of Maverick is intended to provide a single-seat fighter type aircraft with a self-contained day or night Maverick delivery capability. By means of Imaging Infra-Red (IIR) it will be possible to provide aircraft crews with night target acquisition and lock-on capabilities for Maverick the same as those provided by TV Maverick by day.

Maverick-Mk 19:

The 250lb (113kg) Mk 19 warhead is planned for integration with the laser Maverick, but it can also be used with TV or IIR systems. The more powerful warhead provides a 'small skip' kill capability, which is also applicable to other land or sea 'hard' targets. The Mk 19 is a blast/fragmentation type warhead, and its use in Maverick entails an increase in overall length of the missile of about 102 mm and use of a new safe and arming device to allow selectable fuzing.

Blast enhancement and nuclear versions of Maverick have also been considered.

The original total USAF procurement package for 17,000 Maverick missiles initiated in 1968 was completed in November 1975. The following month, Hughes began deliveries under a new contract of a further 6,000, of which 2,000 are the original AGM-65A TV-guided model, and 4,000 are Scene Magnification AGM-65Bs. In addition to the USAF, which deploys Mavericks on F-4D and E, A-7D, and A-10 aircraft, other nations equipped include Israel, Iran, and Saudi Arabia. In April 1976 it was announced that a $20.5 million order for

A-10 ground attack aircraft carries six Maverick air-to-surface missiles

Mavericks to equip Swedish Air Force Viggen aircraft had been signed; these are expected to be in squadron service by 1978.

Laser Maverick moved into the flight test stage in fiscal year 1977 with pilot production projected for December, followed by a fullscale production decision in fiscal year 1978, and a probable total procurement of 4,700 AGM-65C Laser Mavericks. The AGM-76D Imaging Infra-Red (IIR) Maverick engineering development contract is due to be awarded in May 1977, with flight tests following about a year later.

Type: Air-to-surface. Tactical
Configuration: Cylindrical body with rounded nose. Cruciform long-chord delta wings close set to cruciform tail surfaces
Length: 2.46 m
Diameter: 30 cm
Span (Max): 71 cm
Weight: 209 kg
Propulsion: Solid. Thiokol TX-481
Range: 48 km Estimated
Guidance: Automatic TV homing
Warhead: High-explosive. 59 kg shaped charge
Main Contractor: Hughes Aircraft Company
(AGM-65A data)

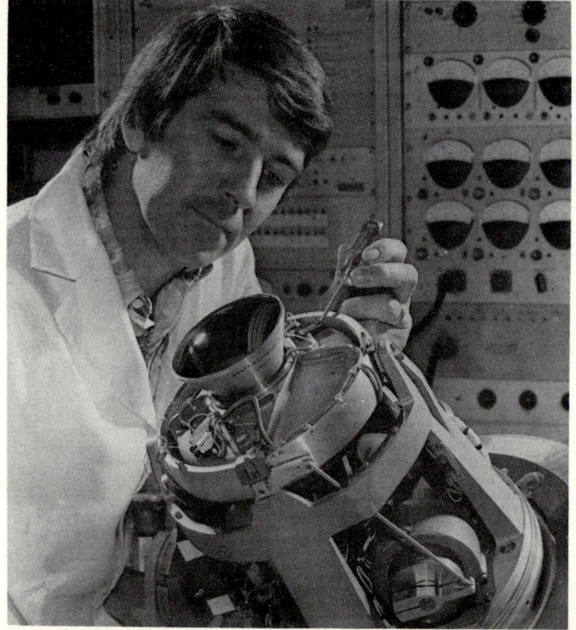

The imaging infra-red (IIR) seeker developed for one of the later versions of Maverick

PAVEWAY

The Paveway range of laser-guided air munitions are similar in concept to the HOBOS (Homing Bomb System) and operationally the two techniques are complementary. Paveway bombs were first used by American forces in Southeast Asia in 1968. Important targets such as bridges, strong-points or tanks are 'marked' by laser from the air or by ground troops and Paveway-armed aircraft are directed by radio to the area. The laser seeker carried on the bomb detects the bright spot of laser light reflected by the target and automatically points itself in the direction of the target. After release from the aircraft the signals from the laser seeker are used to control guidance vanes attached to the bomb to home it onto the target. Kits to convert a number of standard US bombs to laser-guided munitions have been developed.

Four configurations of laser guided bombs (the high and low speed verions of the Mk 82, Mk 83, Mk 84, and the M118) are now operationally deployed with the USN and the USAF. The M117 was initially deployed with the USAF, but is no longer in inventory. The laser-guided Pave Storm cluster munitions (GBU-2) are in development testing.

These eight configurations of laser guided bombs all employing the Texas Instruments guidance kit are compatible with the F-4, F-105, F-111, F-100, F-5, A-6, A-7, A-4, A-1, A-37, B-57, A-10A, F-16, B-52, and AV8A Harrier aircraft. Laser-guided bombs, or LGB kits, have been ordered by Iran and the UK in addition to the US Forces.

Type: Laser-guided bombs
Configuration: Conventional iron bombs and other air-borne munitions modified by addition of laser seekers, guidance computer, and extra control surfaces
Length: 4.27 m
Diameter: 46 cm
Span (Max): 114 cm
Weight: 930 kg
Propulsion: None
Range: Dependent on release height and speed
Guidance: Laser-seeker used in conjunction with air or land-borne laser target marker/designator
Warhead: Various weights of HE and cluster munitions
Main Contractor: Texas Instruments, Inc
(Dimensions and weight are for KMU-351/B model)

Mk 83 Laser-guided bombs on Harrier aircraft

NMC
150529
204

PHOENIX (AIM-54A)

The (AIM-54A) Phoenix missile with its associated AWG-9 fire control radar system (also produced by Hughes) is probably the most sophisticated air-to-air weapon system so far developed. The Phoenix concept was initiated in 1960 and flight testing began in 1965, and testing reached the stage where six missiles were launched almost simultaneously from the same F-14 aircraft and successfully intercepted four out of six targets at a range of more than 80 km. In an earlier test during 1973 a Phoenix launched from an F-14 successfully intercepted a supersonic target drone at a range of 200 km. The AWG-9 radar system which is an essential element in the Phoenix system's ability to engage multiple targets has a total capacity for tracking up to 20 targets simultaneously.

During the cruise phase of missile flight, Phoenix guidance is by semi-active radar homing, the AWG-9 serving as target illuminator. Terminal guidance, reportedly within 16 km of the target, is by active radar. The Phoenix missile incorporates a small doppler radar transmitter which is switched on to provide target illumination during the terminal guidance phase. The complementary I/R target tracker can be used for the direction of missiles at shorter ranges, the cruise phase of flight being unguided until the active radar homer takes over for the terminal phase. In this mode AWG-9 radar need not be used.

First Phoenix units were operational by Spring 1974 and several USN aircraft carriers have F-14 squadrons with Phoenix on board. F-14s of the Imperial Iranian Air Force are to be armed with Phoenix. The Phoenix/AWG-9 combination has been proposed as a naval surface-to-air weapon under the name of Sea Phoenix.

Type: Air-to-air
Configuration: Cylindrical body with slender delta cruciform wings. Four control surfaces at rear end
Length: 3.96 m
Diameter: 38 cm
Span (Max): 91.4 cm
Weight: 380 kg
Propulsion: Solid. Rocketdynne Mk 47 Mod O
Range: 110-160 km Estimated
Guidance: Semi-active radar for 'cruise' phase, active radar terminal homing
Warhead: High-explosive
Main Contractor: Hughes Aircraft Company

Phoenix long-range air-to-air missiles awaiting loading on F-14 aircraft

R.511

(FRANCE)

The R.511 air-to-air interception missile was put into service with French forces in 1958, and somewhat later it was supplied to the Iraqi Air Force. It is not known if it remains operational with either.

Type: Air-to-air
Configuration: Cylindrical body. Two small control surfaces mounted on nose. Rear-mounted wings and ventral rudder
Length: 3.09 m
Diameter: 26 cm
Span (Max): 1 m
Weight: 180 kg
Propulsion: Two-stage solid: 600 kg and 200 kg thrust
Range: 8-10 km
Guidance: Semi-active radar
Warhead: High explosive. Proximity fuzed
Main Contractor: Engins Matra

Matra R.511 air-to-air missile

The all-weather, all-aspect Matra R.530 air-to-air missile entered service in 1963. Two models exist. The infra-red homing version is stated to be capable of successful interceptions from any aspect, relying upon heat from either the jet engine and its exhaust from the rear quarters, and local airframe hot-spots by aerodynamic heating when attacking from the front. This mode of operation is intended mainly for high altitude interception in clear air or at low level beneath cloud. The semi-active radar homing head is produced by Electronique Marcel Dassault under the designation AD26 and this is for use with an I-band target illuminating radar.

The R.530 is in full production and used with the Mirage III, Mirage F1, and Crusader F-8E of the French Naval air arm. In addition to the French Forces, other users include Argentina, Australia, Brazil, Colombia, Lebanon, Pakistan, South Africa, and Venezuela, all of which are equipped with the Mirage. The manufacturer states that fourteen nations have been supplied with this weapon. Production of the R.530 is expected to cease in 1978 and some 4000 have been ordered.

Type: Air-to-air
Configuration: Cylindrical body, with slight taper at fore end to a blunt nose. Cruciform delta planform wings, one pair of which have ailerons. In-line cruciform tail control fins
Length: 3.28 m
Diameter: 26 cm
Span (Max): 110 cm
Weight: 195 kg
Propulsion: Two-stage solid. 8 500 kg static thrust
Range: 18 km
Guidance: Alternative infra-red or semi-active radar homing heads
Warhead: Two alternative types of high explosive heads are available
Main Contractor: Engins Matra

Matra R.530 missiles occupy inner wing pylons and R.550 Magic missiles are carried at wingtips

B37

RB 04E

This air-to-surface missile was developed principally for anti-shipping strikes. The version in current Royal Swedish Air Force service is the Saab 04E which has been developed at Saab-Scania since 1968 under the SAF designation RB 04E. After launch, the missile acts independently of the launch aircraft and is automatically guided to a low-level altitude where target search and acquisition are performed at a high sub-sonic speed. Guidance in the approach phase of an anti-shipping attack with the 04E is by means of a high-quality auto-pilot towards the target area. Precision terminal homing is by an advanced homing head, produced by Philips in Sweden.

The Robot RB 04 series of missiles was initiated by and developed by the Missile Bureau of the Swedish Air Force Board (now the Defence Material Administration — Missiles Directorate). This organisation was responsible for the development of the 04C and 04D versions, but subsequent R & D has been performed by Saab. The 04 missile was first introduced to the Swedish Air Force in 1959/60, and the C and D versions are standard weapon options for the A32A Lansen attack aircraft, which is capable of carrying two of these missiles. Production of the RB 04C ceased in 1964, and the RB 04D entered operational service in 1971. The RB 04E is intended for use with the AJ37 Viggen all-weather attack aircraft, which can carry three missiles. The missile is currently in quantity production, and operational at Swedish Air Force bases.

Type: Air-to-surface. Tactical
Configuration: Circular cross-section fuselage with aft-mounted wings carrying end-plate fins. Cruciform cropped delta canard control surfaces
Length: 4.45 m
Diameter: 50 cm
Span (Max): 2 m
Weight: 600 kg
Propulsion: Solid
Range: Classified
Guidance: Probably active radar
Warhead: High-explosive
Main Contractor: Saab-Scania AB

Swedish Air Force Viggen aircraft carries two RBO4E air-to-surface weapons

SAAB37
-1

RB 05A/B

The Saab RB 05A is a radio command guided, manually controlled tactical air-to-surface missile. It is for use against both sea and land targets but may also be used in certain air-to-air roles. After launching the missile is guided manually by the pilot who visually lines up the missile tracking flares and the target. The control signals, indicated by the pilot through the special control stick, are transmitted over the radio link with the transmitter in the aircraft and the receiver in the missile aft section.

Developed by Saab-Scania on behalf of the Swedish Air Force, the RB 05A is in service with Viggen aircraft. It is also planned to deploy it on various versions of the Saab 105. A new version, designated the RB 05B, was revealed in 1975 to complement the A model. Whereas the 05A is command guided by the pilot via a radio command link throughout the missile's free flight, the 05B is directed to the target by an electro-optical target seeker which the pilot locks on to the selected target before weapon release. After launch the pilot is free to take evasive action.

The future of this development became less certain in 1976 when the Swedish Government decided to procure the TV-guided Maverick from America.

Type: Air-to-surface. Tactical
Configuration: Cylindrical body with pointed nose. Long-chord cruciform wings. Cruciform tail surfaces
Length: 3.6 m
Diameter: 30 cm
Span (Max): 80 cm
Weight: 305 kg
Propulsion: Storable liquid-propellant rocket, type VR-35
Range: Classified
Guidance: Radio command, optical tracking
Warhead: High-explosive. Proximity fuze
Main Contractor: Saab-Scania AB
(Data for RB 05A model)

Viggen (far lest) is armed with RB054, while the RB05B version is seen on this page

RED TOP (UK)

At one time referred to as Firestreak Mk IV, Red Top has a similar configuration and dimensions of the same order, but performance is considerably higher as a result of the application of advances in technology made since the start of Firestreak production. Red Top retains the configuration of four fixed wings and four moving rear control surfaces, the design being further optimised to give operation over a very wide altitude and speed range. Wing and control surface planforms and sections differ from those of Firestreak, and match the unofficially quoted speed of Mach 3. The infra-red guidance system has been further developed to allow target interception from virtually any direction, and a hemispherical nose houses the I/R sensor. Internally, the warhead (31kg) has been moved forward, next to the fuzing system and the control actuators have been located nearer to the surfaces they operate. The power of the internal solid propellant booster rocket motor is also substantially greater than that of Firestreak, giving a range of at least 12 km.

Red Top arms Lightning interceptor aircraft of the RAF and the Kuwaiti and Saudi Arabian air forces.

Type: Air-to-air
Configuration: Cylindrical body with rounded nose. Large cruciform wings mounted near mid-length of missile. Four small triangular control surfaces at tail end
Length: 3.27 m
Diameter: 22.5 cm
Span (Max): 91 cm
Weight: 150 kg Estimated
Propulsion: Solid
Range: 12 km
Guidance: Infra-red
Warhead: High explosive 31 kg
Main Contractor: Hawker Siddeley Dynamics Ltd

Red Top on RAF Lightning

SAAB 372

The Saab 372 is a new air-to-air missile intended primarily for the interceptor version of the Saab Viggen aircraft. Few details have been revealed to date. Infra-red homing guidance will be employed to effect the best compromise between dogfight and intercept missile performance. Range is described as being in the medium class with a short minimum launch range with a good tolerance to launch errors and target manoeuvres. The specific features of the Saab 372 are closley related to the characteristics of the IR target seeker, developed by Saab-Scania, and the associated digital signal processing unit.

The Saab 372 was at the early development phase in May 1975, but studies had preceded this by as much as two years. In July 1975, Saab-Scania was awarded a Swedish Government contract for further development of the complete missile.

Type: Air-to-air
Configuration: Cylindrical body with rounded nose. Cruciform long-chord wings set well aft. Small in-line tail control surfaces
Length: 2.6 m
Diameter: 17.5 m
Span (Max): 62 cm
Weight: 110 kg approximately
Propulsion: Probably solid
Guidance: Infra-red
Main Contractor: Saab-Scania

Saab 372 air-to-air missile

SEA SKUA (CL 834) (UK)

Sea Skua is a new helicopter-launched anti-ship missile system, initially under development for RN Lynx helicopters but expected to be more widely deployed in the future. The operational role of the weapon is to provide long-range self-defence for frigates against missile-carrying fast patrol boats, hydrofoils, and hovercraft having a capability against frigates and larger vessels. It has been stated that the range of the missile itself will be great enough to provide the launching helicopter with a useful degree of 'stand-off' protection from return fire put up by the selected target. This, and the ability of the helicopter to operate at appreciable distances from its parent vessel, are intended to provide protection extending beyond the range of most surface-to-surface anti-ship missiles.

No dimensions have been given officially. A high-explosive warhead weighing about 20 kg is probable. A solid-propellant motor is used, and guidance is by semi-active radar homing, the target being illuminated by the Ferranti Seaspray radar in the Lynx helicopter. After launch, Sea Skua will reduce height to a few metres above wave level for the cruise phase of an attack. A pre-programmed or radio command instruction will bring the missile to target acquisition height for the terminal homing phase. A TRT radio altimeter is fitted for altitude control. A fixed-wing aircraft-launched version is understood to have been studied.

Type: Air-to-surface, anti-ship weapon
Configuration: Long cylindrical body with forepart of larger diameter than rear portion. Rounded nose, Cruciform surfaces near nose and at tail
Length: 2.83 m Provisional
Diameter: 20 cm Provisional
Span (Max): 60 cm Provisional
Weight: 210 kg Provisional
Propulsion: Solid
Range: Classified
Guidance: Semi-active radar homing
Warhead: High-explosive, Approx 20 kg
Main Contractor: British Aircraft Corporation

Sea Skua anti-ship missiles

SHAFRIR

Simplicity was a major design objective of the Shafrir programme which was started in the mid-1960s by Rafael, the Israeli Armament Development Authority. Development was completed by about 1969 and the missile progressively introduced into operational service with the Israeli Air Force. Apart from the firing circuit, no aircraft installed equipment is needed. The missile and its launcher are mounted under the aircraft wing and attached to a specially designed adaptor. This probably contains cooling equipment for the I/R homing head. When a target is detected within the firing range, by the missile's own homing head, an audio signal is given to the pilot as well as a visual warning. From this stage on the pilot is free to use his firing button at any time, after which the guidance system is completely independent of the aircraft.

Official Israeli sources say that Shafrir has been used in combat with a high kill-ratio on numerous occasions. Export sales have been claimed but not identified. Taiwan has been reported as one foreign buyer, other possibilities being Argentina, Peru, and South Africa.

Type: Air-to-air
Configuration: Slim cylindrical body with cruciform wings at rear and four triangular control surfaces near rounded nose
Length: 2.6 m
Diameter: 16 cm
Weight: 93 kg
Propulsion: Solid
Range: Classified
Guidance: Infra-red
Warhead: High explosive, 11 kg
Main Contractor: Rafael Armament Development Authority

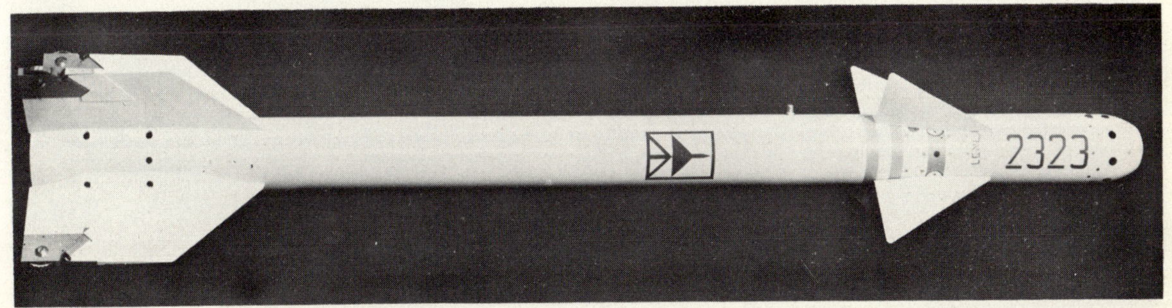

Israeli Shafrir air-to-air missile

REMOVE TO ARM

Shrike has some mechanical affinities with the Sparrow air-to-air missile (which see), largely arising from the urgency which attended development of the Shrike anti-radar missile. Development was initiated by the US Navy in 1961, pilot production started in 1962, full production a year later, and entry into service in 1964.

The principal targets are the radars associated with anti-aircraft gun and missile systems, and there are a number of Shrike homing heads in use to meet specific threats. These each have the appropriate frequency coverage to suit them for engagement of specific types of enemy radars, such as early warning, ground control intercept, and particular surface-to-air missile system radars. Although eventually intended for replacement by HARM (High-Speed Anti-Radiation Missile), Shrike has consistently been procured at annual rates of about 1,000 since its entry into service. In addition to the USN and USAF, quantities of Shrike missiles have been supplied to Israel.

Type: Air-to-surface. Tactical anti-radar missile
Configuration: Slender cylindrical body with pointed nose. Cruciform cropped delta wings near mid-length and small cruciform tail surfaces
Length: 3.05 m
Diameter: 20 cm
Span (Max): 91 cm
Weight: 177 kg
Propulsion: Solid. Rocketdyne Mk 39 Mod 7 or Aerojet Mk 53
Range: 16 km
Guidance: Passive radar homing onto hostile radar transmitters
Warhead: High-explosive
Main Contractors. Texas Instruments and Sperry Rand, Univac

*The Shrike anti-radar missile. (*US Navy*)*

SIDEWINDER (AIM-9) (USA)

Since entering operational service in 1956, the air-to-air AIM-9 Sidewinder has been produced in tens of thousands and supplied to the air forces of over 20 countries. In the course of this long service life (which is continuing) Sidewinder has passed through many stages of development, a high proportion of which have resulted in differing production models. Of necessity, only the most significant aspects can be accommodated in these pages.

The first production model was the AIM-9B, an infra-red homing missile for pursuit engagements. With the exception of the semi-active radar homing AIM-9C, all subsequent versions of Sidewinder have employed IR homing. AIM-9B was widely supplied to NATO and other nations, many of these missiles being updated by the incorporation of an improved guidance and control unit developed by Bodenseewerk Geratetechnik under the designation FGW Mod 2. Missiles converted in this way were known as AIM-9F. Improved American AIM-9Bs became the AIM-9E, this embodying an improved IR head and other modifications to increase low-altitude performance. Also representing an improvement over the AIM-9B, the AIM-9D has a higher thrust motor giving a longer range.

The AIM-9G Sidewinder is similar to earlier versions except for the addition of a vacuum tube and the Sidewinder Expanded Aquisition Mode (SEAM) system, which provides increased lead acquisition capability. The AIM-9H version incorporates SEAM plus solid-state electronics. It became operational with the US Navy in August 1970.

USMC Harrier armed with Sidewinder

The AIM-9J is another improved version of the AIM-9B. This version uses the same seeker as the AIM-9E but has canards of modified design and a high-performance servo system. In March 1972 it was revealed that a number of Sidewinders had been modified for air-to-ground roles, using a laser-seeker guidance system. Such missiles were given the designation Focus (AGM-87A). No details of production contracts have been announced.

The AIM-9L Sidewinder represents the third generation of this widely used air-to-air missile. It differs from its predecessor principally in having an all-aspect attack capability. It is planned for use by both the USN and USAF. An external difference is the altered planform of the canard control fins, the AIM-9L (Fin BSU-32/B) being of double-delta configuration compared with the triangular shape of earlier Sidewinders, and of larger span. Internal changes include a new infra-red homing head using AM-FM conical scan and with increased seeker sensitivity and improved tracking stability. An internal coolant pressure tank (TMU-72/B) is provided for the seeker, this feature enabling the USAF (which does not operate aircraft missile launchers with coolant containers) to use the AIM-9L without modifications to their aircraft. The USN F-4 carries a coolant bottle in the launcher. The AIM-9L also has a significantly enlarged firing envelope, and this feature coupled with the greater manoeuvrability provided by the new control fins enhances the Sidewinder's dogfight capabilities. A new active optical fuze (DU-15/B), allied to the WDU-17/B annular blast fragmentation warhead, increases lethality.

Type: Air-to-air
Configuration: Slender cylindrical body with four large rectangular tail fins and four delta shaped control surfaces near the nose
Length: 2.84 m
Diameter: 12.7 cm
Span (Max): 60.9 cm
Weight: 72 kg
Propulsion: Solid. 2 722 kg' thrust
Range: 3-7 km
Guidance: Infra-red
Warhead: High-explosive
Main Contractor: US Naval Weapons Centre
(AIM-9B Data)

SKY FLASH (XJ521)

Sky Flash, formerly known as UK Sparrow (XJ521), is a new medium range all-weather air-to-air missile for the Royal Air Force. It is based on the Raytheon Sparrow airframe and has the same general configuration and dimensions. It is a semi-active radar guided missile capable of attacking both subsonic and supersonic targets from very low to high altitudes and has an all-round attack capability. It employs a new advanced guidance system developed by Marconi Space and Defence Systems and a new advanced fuze system developed by EMI Electronics Ltd. The auto-pilot and power systems are being updated by Hawker Siddeley Dynamics Ltd, to include solid-state electronics and thermal batteries in place of the existing electronics and power systems to improve reliability. The missile structure will be built by Hawker Siddeley Dynamics who will also carry out assembly and test.

Project definition was completed in 1973 with the development programme following immediately. A contract for full production was awarded in 1975. A number of successful firings have taken place, and trials are continuing to confirm the weapon's capabilities in a hostile electronic warfare environment. In April 1976 it was revealed that first firings of a series of launches from a US Navy F-4J Phantom aircraft against jet target drones at Point Mugu, California, yielded a 100 per cent success rate. Of five launched, three scored direct hits, and two achieved 'lethal passes'. Various types of engagement were involved. Raytheon, manufacturer of the US Sparrow, purchased rights from MSDS for use of the Sky Flash homing head in their own range of Sparrow missiles, and DoD interest has been reported. The Swedish Air Force signed a contract for Sky Flash missiles to arm Viggen aircraft in October 1976. Swedish designation will be RB71.

Launch of Sky Flash from a Phantom aircraft during missile trials

SPARROW (AIM-7E/F)

The Sparrow missile is one of the most widely-used and successful US weapons and it is currently being put to numerous applications in addition to its original role of air-to-air combat. The present production version is the AIM-7E. Semi-active CW radar guidance is employed, with proximity or contact fuzing of the charge. Range is about 25 km.

A version of the AIM-7E Sparrow as used in the NATO Sea Sparrow surface-to-air system bears the designation RIM-7H and differs in having folding fins to reduce the size and weight of the deck launcher/containers. The US Navy Basic Point Defence Missile System uses a version based on the AIM-7D Sparrow.

The latest air-to-air version is the AIM-7F. This has all solid-state electronics for its improved semi-active radar guidance and a larger solid-propellant rocket motor providing increased range. Other features include an increase in reliability; increased launch and attack volume; higher lethality resulting from larger, expanding rod warhead and improved fuze; improved lock-on capability in the presence of look-down clutter; CW and PD illumination radar operating mode compatibility; effective against targets manoeuvring up to 7g.

ECCM performance is limited by the use of conical scanning in the seeker head, but work has started on an improved AIM-7F that will have an inverse monopulse homing head, probably based on the Marconi head for the UK Sparrow, XJ521, to which Raytheon has acquired licence rights. A new active fuze is also planned for this version. The weight is somewhat greater than that of the AIM-7E. The Sparrow missile is also forming the basic vehicle for the USN/USAF Brazo/Pave Arm weapon programme.

Type: Air-to-air, also used in surface-to-air and air-to-surface applications
Configuration: Cylindrical body of slender proportions, with pointed nose. Cruciform triangular wings mid-way along body length; four small triangular fins at rear
Length: 3.66 m
Diameter: 20 cm
Span (Max): 102 cm
Weight: 200 kg
Propulsion: Solid
Range: 25 km approximately
Guidance: Semi-active CW radar
Warhead: High explosive, proximity and contact fuzing
Main Contractor: Raytheon Company

AIM-7F, Sparrow, test firing from USAF F-15 Eagle aircraft

SRAM (AGM-69A)

The Short Range Attack Missile (SRAM) is a supersonic air-to-surface nuclear weapon for the USAF, to complement and soon replace the Hound Dog missile. It is deployed with the B-52G and H, the FB-111, and has been designated for the B-1 strategic bomber. It is designed to attack and neutralise enemy terminal defences, particularly the Soviet surface-to-air missile (SAM) defences. The weapon will have the capability of penetrating terminal defences and striking mission targets while the bombers stand off outside the range of enemy defences. It will also be able to attack enemy SAM and anti-aircraft sites so that bombers can strike primary targets with other SRAM or conventional bombs.

Four basic modes of flight trajectory can be employed: (1) semi-ballistic; (2) altimotor controlled terrain following; (3) pull-up from behind radar screening terrain followed by inertial; and (4) a combination of inertial and terrain following. Lateral deviations in flight profile can also be programmed.

The B-52 carries 20 SRAMs, 12 of which are mounted in two clusters of three missiles on each wing, with eight more carried internally. The FB-111 can carry six missiles, four on individual pivoting pylons under the wings and two internally. The B-1 will have internal stowage for 24 missiles, plus a maximum of eight externally. The rotary launcher to be used with SRAM on B-1 and B-52 aircraft will permit the launching of eight successive missiles at intervals of five seconds.

The initial production of 1,500 SRAMs authorised in 1971 has now been fulfilled and the missiles are deployed with USAF units. Further production for the B-1 bomber depends on Government decisions.

Type: Short Range Attack Missile. Air-to-surface
Configuration: Cylindrical body, tapered at both ends. Three small tail control surfaces. Jettisonable tail cone for external carriage of weapon
Length: 4.27 m, 4.82 m inc. tail cone
Diameter: 45 cm
Weight: 1,000 kg
Propulsion: Two-stage solid
Range: 160 km
Guidance: Programmed inertial
Warhead: Nuclear
Main Contractor: The Boeing Company

SRAM release from FB-111A aircraft. Missile motor ignition had not yet occured

SRAAM

(UK)

SRAAM is a new type of close-combat air-to-air missile, designed to cope with the high accelerations encountered in dog-fight conditions. Its design is such that it can operate down to a very short minimum range, even with the target manoeuvring to its design limits. At the same time, the maximum range is comparable to that of existing air-to-air weapons. Visually aimed, and guided by a passive infra-red homing system, its thrust vector control will enable it to out-manoeuvre all types of aircraft, including crossing targets at very short range. It has a wide aiming tolerance and its high explosive warhead will be detonated by proximity or contact fuzes.

The origins of SRAAM lie in the earlier Hawker Siddeley Taildog project. A UK government definition contract for SRAAM was announced in early 1972. A programme, including air firings, to demonstrate the technology used in SRAAM is in progress.

A surface-launched version for naval air defence has been proposed under the project name Shield.

Type: Short-range air-to-air missile. Dogfight weapon
Configuration: Slim cylindrical body with six folding stabiliser fins at rear end
Length: 2.73 m
Diameter: 16.8 cm
Propulsion: Solid
Range: 400 m – 8 km Provisional
Guidance: Infra-red, missile steering by spoiler vanes in rocket exhaust
Warhead: High explosive
Main Contractor: Hawker Siddeley Dynamics Ltd

Twin launcher tubes and SRAAMs mounted on Hunter aircraft

STANDARD ARM (AGM-78B)

The AGM-78 Standard ARM (Anti-Radiation Missile) is an airborne version of the naval RIM-66A (which see) and which uses its Maxson passive homing head to direct the missile to air defence radars associated with anti-aircraft gun or missile units. It can also be used in a surface launch mode for similar missions against enemy radar installations.

Design studies for Standard ARM and the first development contracts date back to 1966, with flight tests in the following year. Full production began in 1968. The AGM-78B entered service with A-6A and F-105F aircraft in 1968. Production of AGM-78C began in 1970, and a D version will probably arm US Navy EA-6B and E-2C aircraft.

Type: Air-to-surface and surface-to-surface anti-radiation missile
Configuration: Cylindrical body with pointed nose cruciform long-chord wings, with fore portion of reduced span, and cruciform tail surfaces
Length: 4.57 m
Diameter: 30.5 cm
Weight: 816 kg
Propulsion: Dual-thrust solid
Range: 25 km
Guidance: Passive radar, homing onto target's radar emissions
Warhead: High-explosive
Main Contractor: General Dynamics Corporation

Standard ARM (Anti-Radiation Missile).

SUPER 530

(FRANCE)

The Super 530 is a high performance interception air-to-air missile designed as a successor to the Matra R.530. Compared to the latter, the firing range and acquisition are both claimed to be doubled. The Super 530 is intended for use with the next generation of high performance all-weather interceptors and will be able to engage supersonic targets at heights of up to 21,300 m.

The Super 530 development programme was first publicly revealed at the 1971 Paris Air Show, when it was stated that this weapon would arm French Mirage F1 and other interceptor aircraft. Airborne tests of the homing head by EMD were started in September 1972, following initial tests of other sub-systems such as the motor and controls earlier in that year. Flight trials against supersonic targets were due for completion during 1977 and entry into French service was expected in 1978.

Type: Air-to-air
Configuration: Cylindrical body with pointed nose. Cruciform 'long' wings and four tail control surfaces
Length: 3.54 m
Diameter: 26 cm
Span (Max): 90 cm
Weight: 200 kg Provisional
Propulsion: Solid
Range: 29.35 km
Guidance: Semi-active radar
Warhead: High-explosive
Main Contractor: Engins Matra

Test launch of Super 530 air-to-air missile

(USA)

WALLEYE (AGM-62A)

The Walleye weapons family consists of three models of unpowered, guided bombs, designated Walleye Mks I and II and Extended Range, Data Link Walleye. Walleye I has the following main characteristics: length 344 cm, diameter 38.1 cm, wing span 114 cm, weight 449 kg. Conventional, Mk I Mod 0 850lb (385 kg) high explosive warhead.

Walleye II, which has the USN designation Guided Weapon Mk 5, Mod 4, is a much larger missile than the AGM-62A Walleye. Length of Walleye II is 404 cm, body diameter 45.7 cm and weight 1061 kg. A warhead in the 2000 lb (907 kg) class is carried, and the weapon was designed for the destruction of large semi-hard targets such as bridges, air base facilities and ships.

The latest version is Extended Range, Data Link Walleye II (ER/DL WE II) which has the USN designation Mk 13 Mod O. This is essentially a Walleye II fitted with larger wings to extend the glide range and data equipment. The data link was originally intended for single aircraft use of Walleye to enable earlier release (by virtue of the extended glide range) and remote acquisition and lock-on of the Walleye's EO homing system while the launch aircraft had already begun its escape manoeuvre. Experience in SE Asia led to a change in philosophy, by which the data link terminal (and control functions) were transferred to a companion aircraft which followed the Walleye carrier aircraft.

A gyro-stabilised TV camera in the nose of the weapon is aligned with the target prior to launch by the pilot. He is aided in this by a CRT monitor screen, and for target acquisition the fire control radar may be used. When the target has been identified by the pilot, the missile TV camera is locked on and after release the missile guidance and control system uses signals from the TV head to produce control surface movements to direct the bomb to the target. The aircraft is thus free to quit the area immediately after releasing the weapon.

Walleye is extensively deployed with the USN and USAF, but further procurement has probably been deferred in favour of the GBU-15(V) Modular Glide Weapon System, described earlier.

Walleye guided bomb. (US Navy)

INDEX

N

O

P

R

S

T

U

V

W

X

Y